H. H. (Hans Herman) Behr

**Flora of the Vicinity of San Francisco**

H. H. (Hans Herman) Behr

**Flora of the Vicinity of San Francisco**

ISBN/EAN: 9783744666800

Printed in Europe, USA, Canada, Australia, Japan

Cover: Foto ©berggeist007 / pixelio.de

More available books at **www.hansebooks.com**

# FLORA

OF THE

Vicinity of San Francisco.

By

H. H. BEHR, M. D.

SAN FRANCISCO, CAL.,

1888.

# PREFACE.

The territory to which this book applies extends from Sonoma to Santa Clara, and from Niles to the Pacific Ocean.

Since the year 1884, when the Synopsis of Genera was published, there have been added several genera, which up to that time had escaped observation. Continued investigations will lead perhaps to the discovery of some more; certainly, they will lead to the addition of species hitherto unobserved in the territory. I have followed as much as possible the admirable work of Asa Gray and Sereno Watson "Botany of California," and from their views I have deviated only when opportunities for studying the plants from specimens growing on the spot afforded to me ample grounds for a view differing from that expressed by recognized authorities.

As to many of the species characterized in this book, there exists still a considerable degree of uncertainty. Many of our Californian species split into numerous variations, which mingle frequently with variations of related, equally variable species. Some of these variations owe their existence to hybridization; and this circumstance is probably the reason why several species described and characterized by different authors have not been found again. In annuals such spurious species will only reappear occasionally.

Questionable types can be investigated only by cultivation. Up to this time California does not possess a botanical garden or experimental grounds where such questions could be definitely settled. Therefore the solution of many problems has been left entirely to enthusiasts who utilize their leisure hours in the cultivation of doubtful species, frequently under great difficulties.

Owing to these circumstances, there is scarcely an author writing on our Flora who has not been compelled to frequently change or correct his previous views, and I hope the present author will be excused if in future times investigations shall lead him to add to, or to retract, some of his statements.

H. H. BEHR, M. D.

# LINNÆAN KEY.

## MONANDRIA.

I. **Monogynia.**
1. Acæna.—Calyx lobes valvate. Style terminal. Akene enclosed in the calyx.
2. Hippuris.—Calyx superior. Aquatic.

II. **Digynia, Trigynia and Tetragynia.**
1. Acæna.—Calyx lobes valvate. Akenes enclosed in the calyx. Styles of akenes terminal.
2. Callitriche.—Fruit a 4-coccous schizocarp. Aquatic.
3. Chenopodium.—Perigonium 3-5-parted. Fruit a utricle enclosed in the persistent perigonium.
4. Festuca.—Glumaceous.
5. Cyperus.—Glumaceous, trigynous.

## DIANDRIA.

I. **Monogynia.**

A. *Floral parts incomplete, inferior.*
1. Salicornia.—Perigonium gamosepalous, opening by a slit and immersed in an excavation of the rachis.

2. Lemna.—Perigonium compressed, growing out of the margin of a thallus. Aquatic.
3. Scirpus.—Glumaceous. Spikelets imbricate from all sides. Inferior glumæ (1-2) sterile, but not smaller than the rest. Perigonial bristles not exserted. Fruit, a nutlet, apiculate by the persistent base of the style. Aquatic.
4. Fraxinus.—Ovary 2-celled, laterally compressed. Fruit, a samara. Tree.
5. Acæna.—Perigonium lobes valvate. Fruit an akene enclosed in the perigonium. Style terminal.
6. Alchemilla.—Perigonium urceolate, bracteolate, not spinose. Akene enclosed in the perigonium. Style lateral.

B. *Floral parts complete, inferior, regular.*

1. Fraxinus.—Fruit, a samara. Tree.
2. Lepidium.—Fruit, a silicula, laterally compressed.

C. *Floral parts complete, inferior, irregular.*

    a. *Ovary entire.*

1. Veronica.—Calyx herbaceous 4-parted. Ovary 2-celled. Fruit, an emarginate capsule. Leaves opposite.

2. Synthyris.—Calyx herbaceous, 4-parted. Ovary 2-celled.
3. Cordylanthus.—Calyx spathaceous, 1-sepalous. Ovary 2-celled.
4. Utricularia.—Ovary 1-celled, centrospermous.

 b. *Ovary divided into 4 partitions.*

1. Lycopus.—Anthers 2-celled. Cells parallel. Aquatic.
2. Salvia.—Anthers 1-celled. Filament with a lateral branch, geniculate.
3. Audibertia.—Anthers 1-celled. Filament without lateral branch or spur.
4. Acanthomintha.—Stamens ascending. Anthers 2-celled. Cells divaricate. A pair of staminodia indicating the place of the upper pair of stamens.

II. **Digynia, Trigynia and Tetragynia.**

1. Acæna.—Calyx lobes valvate, tube spinose. Fruit, akenes enclosed in the calyx. Styles terminal.
2. Alchemilla.—Calyx urceolate, bracteolate, not spinose. Fruit, akenes enclosed in the calyx.
3. Chenopodium.—Perigonium 3–5-parted, not appendiculate. Fruit a compressed utricle, inclosed in the perigonium.
4. Festuca.—Glumaceous.

## TRIANDRIA.

I. **Monogynia.**

A. *Floral parts complete, superior.*

1. Plectritis.—Limb of calyx obsolete. Corolla gamopetalous, calcarate.

B. *Floral parts complete, inferior.*

1. Montia.—Calyx 2-sepalous. Corolla cleft on one side, 5-parted.

C. *Floral parts incomplete.*

   a. *Perigonium petaloid.*

1. Iris.—Perigonium 6-parted, superior, alternate lobes reflexed.
2. Scoliopus.—Perigonium 6-sepalous inferior, alternate lobes reflexed.
3. Brodiæa.—Perigonium 6-parted. inferior. Stamens alternating with staminodia.

   b. *Perigonium herbaceous.*

1. Acæna.—Calyx lobes valvate, tube spinose. Fruit an akene enclosed in the calyx. Style terminal.
2. Chenopodium.—Perigonium 3-5-parted, not appendiculate. Fruit a compressed utricle enclosed in the perigonium.

C. *Floral parts glumaceous.*
  aa. *Spikelets distichous.*
  1. Cyperus.—Glumæ ∞, one or the two lowermost smaller and empty.
    bb. *Spikelets imbricate on all sides.*
  1. Eleocharis.—Perigonial bristles not exserted. Caryopsis apiculate by the persistent articulate base of the style.
  2. Fimbristylis.—Perigonial bristles not exserted (sometimes wanting). Caryopsis apiculate by the persistent continuous base of the style. Aquatic.
  3. Scirpus.—Perigonial bristles not exserted (sometimes wanting altogether). Caryopsis apiculate by the persistent continuous base of the style. Aquatic.
  4. Eriophorum. — Perigonial bristles exserted, considerably longer than the glumæ. Aquatic.

Observation.—All grasses which are monogynous are put in Triandria Digynia.

II. **Digynia.**

A. *Flowers glumaceous.*
  a. *Spikelets single 1-flowered, imbedded in the notches of the rachis, each covered by its glume.*

1*

I. **Lepturus.**

 b. *Spikelets all sessile on alternating notches of the rachis.*

  1. Lolium.—Terminal spikelet with 2 glumes, lateral spikelets but one glume.
  2. Hordeum.—Spikelets 3 at each notch of the rachis, 1-flowered. Fruit a caryopsis wrapped and grown together with the paleæ.
  3. Elymus.—Spikelets 3 at each notch of the rachis, 2—∞-flowered.
  4. Gymnostichum.—Spikelets 1-3 at each notch of the rachis. Glumæ rudimentary or 0.
  5. Triticum.—Spikelet 1 at each notch of the rachis. Glumæ 2, placed right and left (not dorsal and ventral.)

 c. *Spikelets pedicillate, 1-3-flowered, but only 1 flower regularly developed.*

aa. *Spikelets dorsally compressed.*

  1. Panicum.—Glumæ 3, the lowermost very small. Involucre 0.
  2. Setaria.—Glumæ 3, the lowermost small; bristly involucre below the spikelet.

bb. *Spikelets laterally compressed. Glumæ 3. The two inferior florets rudimentary, terminal floret only developed.*

1. Phalaris.—The 2 inferior florets reduced to 2 awnless paleæ.
2. Anthoxanthum.—The two inferior florets reduced to 2 aristate paleæ larger than the developed terminal floret.

cc. *Spikelets not dorsally compressed. Glumæ 2. When a second floret is present it is rudimentary and terminal.*

aaa. *Stigma filamentose, exserted on the apex of the spikelet.*

1. Alopecurus.—Floret consisting of a single palea.
2. Phleum.—Glumæ carinate, longer than the 2 paleæ of the floret.
3. Spartina.—Glumæ carinate, the inferior shorter than the floret. Paleæ unequal, the superior longer.

bbb. *Stigma on an elongated style, aspergilliform and exserted below the apex of the floret.*

1. Cynodon.—Glumæ small, spreading. Superior palea linear, concave; inferior palea laterally compressed, oval, chartaceous.

ccc. *Stigma plumose, exserted from the base of the floret. Style very short.*

1. Polypogon.—Glumæ obtuse, but aristate. Paleæ membranaceous.

2. Agrostis.—Glumæ acute, the inferior one larger. Paleæ membranaceous.
3. Calamagrostis.—Glumæ acute, the inferior one larger. Paleæ surrounded by silky hair at the base.
4. Stipa.—Glumæ acute, sometimes aristate. Paleæ cartilaginous, the inferior one cylindrically involute. Awn strong, articulate at the base, persistent, developed from the apex of the palea.
5. Gastridium.—Glumæ acute, very much compressed, ventricose at their base. Paleæ membranaceous.

d. *Spikelets pedicillate, 2-∞-flowered, the uppermost floret frequently rudimentary.*

aa. *Stigma aspergilliform, exserted below the apex of the floret.*

1. Phragmites.—Spikelets ∞-flowered, lowermost floret ♂ or neuter, the other florets ☿ and pilose.
2. Hierochloa.—Spikelets 3-flowered, the two inferior ones ♂ and triandrous, the terminal one ☿ and diandrous.

bb. *Stigma plumose, exserted near the base of the floret.*

1. Arrhenatherum.—Spikelets 2-flowered; inferior floret ♂, aristate; superior floret ☿, awnless.

2. Holcus.—Spikelets 2-flowered, inferior floret ☿, awnless; superior floret ♂, aristate.
3. Aira.—Spikelets 2-flowered, both ☿; inferior palea 4-dentate at its apex, aristate from its base.
4. Avena.—Spikelets 2-∞-flowered. Florets ☿. Inferior palea 2-dentate at its apex, aristate from its median nerve.
5. Danthonia.—Spikelets 2-∞-flowered. Florets ☿. Inferior palea 2-cleft, aristate from the fissure.
6. Melica.—Spikelets 2-∞-flowered, florets awnless, the inferior or 2 inferior, ☿; the rest rudimentary.
7. Kœleria.—Glumæ and paleæ membranaceous. Spikelets 2-∞-flowered. Florets lanceolate, compressed, carinate; inferior palea mucronate-aristate.
8. Dactylis.—Glumæ and paleæ herbaceous. Spikelets 3-∞-flowered. Florets ovate, carinate. Inferior palea asymmetric, mucronate-aristate.
9. Poa.—Spikelets 2-∞-flowered. Florets ovate, carinate, awnless, deciduous with the joint of the axis.
10. Eragrostis.—Spikelets 2-∞-flowered. Florets ovate, carinate, awnless; superior palea and axis persistent.

11. Glyceria. Spikelets 2, ∞-flowered. Florets ovate, convex, awnless. Aquatic grass.
12. Lophochlæna. — Spikelets ∞-flowered, compressed, Glumæ membranaceous. Inferior palea 2-lobed, aristate from the mid-nerve. Superior palea 2-carinate with a pair of dentate appendages.
13. Briza.—Spikelets 2- ∞-flowered, densely distichous. Florets awnless. Inferior palea inflated, with auriculate cordate base.
14. Festuca.—Spikelets 2- ∞-flowered. Florets lanceolate, convex. Superior palea ciliate. Style inserted on the apex of the ovary.
15. Bromus.—Spikelets ∞-flowered. Florets lanceolate. Style inserted below the apex of the ovary on its **ventral** side.

B. *Flowers not glumaceous.*
1. Pentacæna.—Calyx 2-seriate, persistent. External sepals 3, cucullate at the apex, ending in a spine; internal sepals 2, mucronate. Fruit a utricle.

### III. Trigynia.
1. Tillæa.—Calyx 3-sepalous. Corolla 3-petalous. Fruit 3 follicles, each 2-seeded. Aquatic.

2. Lastarriæa.—Perigonium tubular, coriaceous, 6-dentate. Ovary 1. Fruit an akene. Involucre 0.
3. Chorizanthe.—Involucre 1-3 flowered, coriaceous, 3-6 angular. Ovary 1. Fruit, an akene.
4. Tricerastes (Datisca).—Calyx superior Petals 0. Anthers extrorse. Ovary 3-quetrous. Styles 3, each 2-cleft. Ovary 1-celled. Placentæ parietal.
5. Cyperus.—Glumaceous. Spiklets distichous.
6. Eleocharis.—Glumaceous. Spiklets imbricate. Perigonial bristles not exserted. Caryopsis apiculate by the persistent articulate base of the style.
7. Fimbristylis.—Glumaceous. Spikelets imbricate. Perigonial bristles not exserted. Caryopsis apiculate. Style articulate above its base. Aquatic.
8. Scirpus.—Glumaceous. Spiklets imbricate. Perigonial bristles not exserted. Caryopsis apiculate by the persistent continuous base of the style. Aquatic.
9. Eriophorum.—Glumaceous. Spikelets imbricate. Perigonial bristles exserted, considerably longer than the glumæ. Aquatic.

## TETRANDRIA.

I. **Monogynia and Digynia.**

A. *Flowers complete.*

    a. *Calyx double; the interior calyx adnate to the ovary. Corolla gamopetalous.*

1. Dipsacus.

    b. *Calyx simple. Corolla gamopetalous, inferior.*

1. Plantago.—Corolla-limb regular, 4-parted. Stigma filamentose. Capsule circumscissile.
2. Gentiana.—Ovary 1. Placentæ 2. Style short. Stigmas 2, flat and thin. Capsule septicidal. Seeds minute.
3. Microcala (Cicendia).—Ovary 1. Placentæ 2. Style filiform. Stigma peltate. Seeds minute.

    c. *Corolla gamopetalous, superior.*

1. Galium.—Corolla rotate. Fruit a 2-coccous schizocarp.

    d. *Corolla 4-petalous, inferior.*

1. Euonymus.—Petals inserted on a glandular disc. Stamens inserted on the same disc, and alternating with them. Fruit a capsule. Shrub.

2. Ptelea.—Stamens alternating with the petals. Ovary on a convex disc. 2-celled. Cells 2-ovulate. Fruit a 2-seeded samara. Shrub.

    e. *Corolla 4-petalous, superior.*

1. Cornus.—Calyx 4-parted. Fruit a drupe. Tree. Shrub.
2. Eucharidium.—Calyx tube extended beyond the ovary, its lobes reflexed. Petals unguiculate, lobed. Fruit a capsule.
3. Gayophytum.—Calyx tube not extended beyond the ovary. Anthers versatile. Stamens alternating with staminodia, which are opposite to the petals. Fruit a capsule.

B. *Flowers incomplete.*

    a. *Perigonium inferior.*

1. Majanthemum.—Perigonium corolline, 4-parted, rotate.
2. Acæna.—Calyx lobes valvate, tube spinose. Ovaries 1-2. Style or styles terminal. Fruit 1 or 2 akenes, enclosed in the calyx tube.
3. Alchemilla.—Perigonium calycine; limb 8-cleft. Stamens inserted on an annular throat (faux).

4. Sanguisorba.—**Perigonium** calycine; limb 4-cleft, colored. Stamens inserted on an annular throat (faux).

5. Negundo.—Perigonium very small, 4-dentate. Fruit a 2-seeded samara. Tree. Shrub.

## II. Tetragynia.

1. Tillæa.—Calyx 4-sepalous. Corolla 4-petalous. Follicles 4, each 2-seeded.
2. Sagina.—Calyx 4-sepalous. Corolla 4-petalous. Capsule 4-valved. Seeds reniform.
3. Cuscuta.—Corolla gamopetalous, 4-cleft. Ovary 1; ovules few, inserted on the base of the ovary. Capsule circumscissile. Parasitic.
4. Potamogeton. — Perigonium 4-parted. Fruit, 4 sessile drupes. Aquatic.

### PENTANDRIA.

**I. Monogynia.**

A. *Floral parts gamopetalous, inferior.*

   a. *Fruit 4 nutlets, each 1–2 seeded.*

   aa. *Ovary with 4 sutures finally separating into 4 nutlets.*

1. Heliotropium.—Corolla funnel-shaped, throat naked, limb plicate.

bb.—*Ovary 4-parted, nutlets ventrally adherent to the persistent style.*

1. Echinospermum. — Nutlets 3-quetrous, glochidiate on their edges.
2. Cynoglossum.—Nutlets compressed, spinose on their entire dorsal surface.
3. Pectocarya.—Nutlets in pairs, divergent, carinate, carina dentate.

   cc. *Ovary deeply 4-parted, nutlets excavated on their ventral side inserted on a pyramidal disc.*

1. Amsinckia.—Nutlets coriaceous, inserted above their base. Cotyledons 2-cleft.

   dd. *Ovary deeply 4-parted, nutlets inserted on a flat disc.*

1. Krynitzkia (Eritrichium).—Nutlets ventrally attached from near their base to a columnar prominence inserted on the otherwise flat disc.
2. Plagiobothrys [Eritrichium]. — Nutlets attached near the middle of their ventral face to a columnar prominence, inserted on the otherwise flat disc.

   b. *Fruit a 1-celled capsule with free central placentation.*

1. Dodecatheon. — Corolla partitions reflexed.

2. Anagallis.—Corolla rotate, 5-parted, capsule circumscissile.

 c. *Ovary 1-celled with 2 parietal placentæ.*

1. Menyanthes. — Corolla funnel-shaped. Stigma undivided. Capsule 2-valved, valves bearing the seeds in their middle; placentæ not detached from the exocarp. Aquatic.
2. Nemophila.—Calyx with reflexed sinuses. Stamens shorter than the corolla. Stigmas 2. Capsule 2-valved. Placentæ detached from the exocarp.
3. Ellisia.—Calyx without reflexed sinuses. Stamens shorter than the corolla. Stigmas 2. Capsule .2-valved. Placentæ detached from the exocarp.
4. Romanzoffia. — Calyx without reflexed sinuses. Corolla funnel-shaped. Stigma 1, undivided. Capsule loculicidal. Seeds $\infty$.

 d. *Capsule 2-$\infty$-celled.*

1. Erythræa.—Corolla funnel-salver-shaped. Anthers finally spiral. Seeds minute.
2. Gentiana.—Ovary 1. Placentæ 2. Style short. Stigmas 2. Capsule septicidal. Seeds minute.

3. Polemonium.—Corolla rotate. Stigmas 3. Filaments inserted at equal heights on the tube of the corolla, ascending; anthers incumbent.
4. Gilia.—Corolla rotate—funnel-shaped. Stigmas 3. Filaments straight, inserted at equal heights; anthers incumbent.
5. Collomia.—Corolla salver-shaped. Stigmas 3. Filaments inserted at unequal heights and exserted.
6. Convolvulus.—Corolla funnel-shaped, plicate, 5-angular. Cells of ovary 2-ovulate. Calyx marcescent.
7. Datura.—Corolla funnel-shaped, plicate, 5-angular. Calyx transversely deciduous.
8. Nicotiana.—Corolla funnel-shaped, plicate, 5-angular. Calyx persistent. Cells of ovary $\infty$-ovulate.
9. Rhododendron.—Corolla campanulate. Stamens inserted on a disc, not on the corolla. Cells of the anthers opening by pores. Shrubs.
10. Pentstemon.—Corolla irregular, funnel-shaped. Stamens of unequal length, one of them sterile. Capsule 2-celled $\infty$-ovulate.

e. *Fruit baccate.*

1. Solanum. — Corolla rotate. Anthers opening by apical pores.

B. *Floral parts gamopetalous, superior.*
   a. *Fruit capsular.*

1. Samolus. — Staminodia 5, alternating with the stamens. Ovary 1-celled.
2. Campanula.— Corolla 5-lobed. Ovary flat at its apex. Capsule turbinate dehiscent by lateral pores.
3. Heterocodon (Campanula).—Corolla 5-lobed. Ovary flat at its apex. Capsule turbinate, laterally but irregularly dehiscent.
4. Specularia.— Corolla rotate. Capsule elongated, prismatic, dehiscent laterally by valves.
5. Githopsis.—Corolla tubular. Capsule clavate, dehiscent at its apex by the falling off of the base of the style.
6. Downingia. — Corolla irregular. Filaments and anthers united into a tube.

   b. *Fruit baccate.*

1. Lonicera.—Corolla irregular. Ovary 2-3-celled, ∞-seeded.
2. Symphoricarpus.—Ovary 4-celled. Berry 2-seeded. Shrubs.

c. *Floral parts polypetalous, inferior.*

aa. *Floral parts irregular.*

1. Viola.—Calyx 5-sepalous. Corolla 5-petalous. Capsule 3-valved, placentæ 3; each ∞-seeded.
2. Æsculus.—Calyx 5-dentate. Corolla 4–5-petalous. Capsule 3-valved, 1-seeded. Tree.

bb. *Floral parts regular.*

1. Acer.—Fruit a 2-seeded samara. Trees. Shrubs.
2. Vitis.—Petals coherent at the apex, and caducous at the base, inserted on a disc. Fruit baccate.
3. Euonymus.—Petals 5, inserted on the calyx. Stamens alternating with the petals, inserted on a glandular disc. Fruit capsular. Shrubs.
4. Rhamnus.—Petals 4–5 with short claws, inserted on the calyx. Stamens opposite to the petals. Fruit a drupe with several pyrenæ. Trees. Shrubs.
5. Ceanothus.—Calyx and disc adnate to the base of the ovary. Stamens opposite to the petals. Petals with long claws cucullate. Fruit a capsule, 3-celled; cells 1-seeded. Capsule septicidal. Trees. Shrubs.

6. Ptelea.—Stamens alternating with the petals. Ovary on a convex disc, 2-celled; cells 2-ovulate. Fruit a 2-seeded samara Shrubs.
7. Erodium.—Carpidia 5, verticillate round a columnar axis. Styles connate at their apex.
8. Pentacœna.—Sepals 2-seriate, persistent; external sepals 3, with cucullate apex terminating in a spine, internal sepals, mucronate. Ovary 1-ovulate. Style 2-cleft. Fruit a utricle.
9. Frankenia.—Calyx tubular, 4–5 lobed. Petals 4 or 5, unguiculate. Capsule 1-celled, dehiscent by valves. Ovules few, inserted on parietal placentæ.

d. *Floral parts polypetalous, superior.*

1. Ribes.—Petals unguiculate and with the stamens inserted on the calyx limb. Fruit baccate, ∞-seeded. Shrubs.

e. *Floral parts incomplete.*

1. Glaux.—Perigonium campanulate, 5-lobed, petaloid. Ovary 5-valved.
2. Polygonum.—Perigonium 5-sepalous, persistent. Stigmas 2–3, capitulate. Fruit an akene enclosed in the persistent perigonium.

3. Acæna.—Calyx lobes valvate, tube spinose. Style terminal. Akene enclosed in the calyx tube.
4. Negundo.—Perigonium small, 5-dentate. Fruit a samara. Tree.
5. Abronia.—Flowers capitulate. Involucre ∞-phyllous. Perigonium gamosepalous, salver-shaped.

## II. Digynia.

A. *Flowers incomplete.*

1. Suæda (Schoberia).—Perigonium 5-parted, not appendiculate. Embryo spiral.
2. Chenopodium.—Perigonium 5-parted, not appendiculate. Embryo annular.
3. Acæna.—Calyx lobes valvate, tube spinose. Fruit, 2 akenes, enclosed in the calyx tube.

B. *Flowers complete, gamopetalous; floral parts inferior.*

    a. *Ovaries 2. Seeds* ∞.

1. Apocynum.—Ovaries distinct at the base but united by their stigma. Five scales in the tube of the corolla opposite to the corolla lobes.

b. *Ovary 1, with two opposite placentæ.*

1. Gentiana.—Lobes of the corolla without nectaria. Capsule ∞-seeded. Seeds minute.
2. Nemophila.—Calyx with reflexed sinuses. Stamens shorter than the corolla. Capsule 2-valved. Placentæ detached from the exocarp.
3. Ellisia.—Calyx without reflexed sinuses. Stamens shorter than the corolla. Capsule 2-valved. Placentæ detached from the exocarp.
4. Phacelia.—Corolla deciduous. Stamens exserted, equal. Capsule loculicidal.
5. Emmenanthe.—Corolla persistent. Stamens exserted, equal. Capsule loculicidal.
6. Eriodictyon.—Capsule 4-valved (loculicidal and septicidal). Shrub.

c. *Ovary 1. Seeds few. Placentæ on the base of the ovary.*

1. Cuscuta.—Capsule circumscissile. Parasite.

C. *Corolla 5-petaled. Floral parts inferior.*

1. Heuchera.—Ovary 1-celled. Capsule ∞-ovulate, dehiscent by valves. Pla-

centræ parietal at the margins of the valves.
2. Pentacæna.—Calyx 2-seriate. Petals minute. Ovary 1-ovulate. Fruit a utricle.

D. *Corolla 5-petaled. Floral parts superior.*
   a. *Flowers in a simple umbel.*
   1. Panax.—Fruit baccate.
   2. Hydrocotyle. — Petals entire, acute. Akenes laterally compressed. Costæ filiform.
   3. Bowlesia.—Petals obtuse. Akenes ovate, turgid, without costæ, pubescent.
   4. Eryngium. — Petals erect, connivent, emarginate at their apex. Akenes terete, obovate, without costæ but squamate and tuberculate.
   5. Sanicula. Petals erect, connivent, emarginate at their apex. Akenes spinose.
   6. Panax. Fruit baccate.
   b. *Flowers in a compound umbel.*
   aa. *Endosperm not excavated on its ventral surface (Orthospermæ). Costæ 5, filiform; secondary costæ 0. Akene laterally compressed.*

aaa. *Petals entire.*
   1. Apium. -Petals stellate, rounded, with involute apex.

bbb. *Petals obcordate with a small inflexed lobe. Calyx limb obliterated.*

1. Carum.—Petals regular. Styles reclined. Akenes oblong. Costæ filiform. Oil tubes solitary.
2. Pimpinella. — Petals regular. Styles thin, divergent. Costæ fiiliform. Oil tubes 3 in each vallecula.

ccc. *Petals obcordate with a small inflexed lobe. Calyx limb 5-dentate.*

1. Cicuta.—Endosperm convex. Fruit contracted. Oil tubes solitary.
2. Sium.—Fruit contracted. Oil tubes superficial, 3 in each vallecula.
3. Berula.—Fruit contracted. Oil tubes covered by a thickened pericarp, 3 in each vallecula.

bb. *Endosperm not excavated on its ventral surface (Orthospermæ). Costæ 5, secondary costæ 0. Akenes terete or subterete.*

1. Angelica.—Petals not obcordate, lateral costæ extended into wings.
2. Selinum.—Calyx limb obliterated. Petals obcordate. Oil tubes solitary. Lateral costæ much broader than the dorsal.
3. Œnanthe.—Calyx limb 5-dentate. Styles erect. Oil tubes solitary. Carpophore adnate.

cc. *Endosperm not excavated on its ventral surface (Orthospermæ). Costæ 5, filiform, secondary costæ 0. Akenes dorsally compressed, lenticular.*

1. Heracleum.—Petals obcordate. Oil tubes clavate. Lateral costæ of the two akenes touching each other and more distant from the 3 dorsal costæ than these amongst themselves.
2. Peucedanum.—Lateral and dorsal costæ equally distant. Oil tubes superficial.

dd. *Primary costæ 5, secondary costæ 4.*

1. Daucus.—Primary costæ fiiliform, bristly. Secondary costæ spinose.

ee. *Endosperm concave on its ventral surface by a longitudinal furrow (Cœlospermæ).*

1. Caucalis.—Primary costæ spinose; secondary more prominent than the primary.
2. Conium.—Akenes laterally compressed. Costæ undulate, crenulate; secondary costæ 0. Oil tubes 0.
3. Deweya.—Akenes laterally compressed. Costæ prominent. Secondary costæ 0. Oil tubes 2–3 in each vallecula.
4. Osmorrhiza.—Akenes linear, angulate. Carpophore persistent.

C. *Flowers not umbellate.*
1. Boykinia.—Ovary 2-celled. Capsule ∞-seeded.
2. Heuchera.—Ovary 1-celled. Capsule ∞-seeded, dehiscent by valves.
3. Ribes.—Petals and stamens inserted on the calyx limb. Petals unguiculate. Fruit baccate, ∞-seeded. Shrubs.

**III. Trigynia.**

A. *Floral parts complete, inferior.*
1. Rhus.—Fruit drupaceous, 1-pyrenous. Pyrena 1-seeded. Shrubs.
2. Calandrinia.—Sepals 2, persistent. Petals 5. Capsule 3-valved 2-celled. Placenta central, ∞-seeded. Succulent.
3. Claytonia.—Sepals 2, persistent. Petals 5. Capsule 3-valved, 1-celled. Placenta central, few-seeded. Succulent.
4. Linum.—Sepals 5. Capsule 3-celled. Cells 2-ovulate.
5. Alsine.—Sepals 5. Petals 5. Capsule 1-celled, 3-valved. Placenta central.

B. *Floral parts complete, superior.*
1. Sambucus. Corolla rotate. Berry 3-seeded (Drupe 3-pyrenous.) Tree. Shrub.
2. Whipplea.—Ovary 3-celled, cells 1-ovulate. Fruit a septicidal capsule.

## IV. Pentagynia.

1. Aralia.—Fruit a 5-pyrenous drupe. Pyrenæ 1-seeded.
2. Tillæa.—Calyx 5-parted. Petals 5. Fruit 5-follicles. Aquatic.
3. Linum.—Sepals 5. Capsule 4–5-celled, cells 2-ovulate.
4. Statice.—Calyx limb scarious. Ovary 1-celled, 1-ovulate. Styles not plumose.
5. Armeria.—Calyx limb scarious. Ovary 1-celled, 1-ovulate. Styles plumose.
6. Spergula.—Calyx 5-sepalous. Petals 5, entire. Capsule 1-celled, 5-valved. Seeds winged.
7. Sagina.—Calyx 5-sepalous. Capsule 1-celled, 4–5-valved. Seeds reniform.

## V. Polygynia.

1. Myosurus.—Claws of petals longer than the lamina. Carpidia ∞. Akenes 1-seeded.

## HEXANDRIA.

### I. Monogynia.

A. *Floral parts complete.*

1. Berberis.—Petals 6, inferior. Fruit a berry. Shrub.
2. Vancouveria.—Petals 6, inferior. Fruit a follicle.

3. Trillium.—Sepals 3. Petals 3. Stigmas 3, almost sessile.
4. Lythrum.—Calyx tubular, 8–12-dentate. Petals 4—8.
5. Trientalis.—Calyx 6–7 cleft. Corolla 6–8 parted, spreading. Capsule 6–8-valved, 1-celled. Placenta central.
6. Frankenia.—Calyx tubular, 5-cleft. Petals 5, unguiculate. Capsule 1-celled, dehiscent by valves. Ovules few, inserted on parietal placentæ.

B. *Floral parts incomplete. Perigonium corolline, gamosepalous.*

1. Muilla (Hesperoscordum) Perigonium deeply 6-parted, campanulate. Stamens inserted on the base of the perigonium.
2. Brodiæa.—Perigonium 6-lobed, funnel-shaped. Stamens inserted on the throat, alternately sterile.
3. Triteleia.—Perigonium 6-lobed, salver-shaped. Stamens inserted alternately on the throat and on the middle of the tube,

C. *Floral parts incomplete. Perigonium corolline, 6-sepalous.*

    *a. Style 3-cleft, or 3 sessile stigmas.*

1. Fritillaria.—Sepals of perigonium with a nectarium at the base.

2. Calochortus.—Alternate sepals of different shape; interior sepals larger. Stigmas sessile. Anthers basi-fixed. Capsule 3-celled.

3. Trillium.—External sepals herbaceous, internal corolline. Stigmas almost sessile. Fruit a berry.

    b. *Style entire or 0. Stigma entire or 3-lobed.*

    aa. *Anthers basi-fixed.*

1. Prosartes.—Fruit baccate; testa of the seeds thin, membranaceous.
2. Clintonia.—Fruit baccate; testa of the seeds crustaceous.

    bb. *Anthers versatile.*

1. Lilium.—Sepals with a nectariferous longitudinal groove.
2. Allium.—Nectariferous grooves 0. Flowers umbellate, before spreading enclosed in a spathe.
3. Cyanotris (Camassia).—Sepals 6, 5 ascending, 1 deflexed. Stamens ascending. Style declined, its base persistent. Fruit a loculicidal capsule, several seeded.

4. Chlorogalum. Sepals spreading, ligulate. Stamens spreading. Style deciduous. Fruit a loculicidal capsule Each cell 2-seeded.
5. Smilacina.—Style short, 3-lobed, persistent. Fruit baccate, few-seeded.

D. *Floral parts incomplete, Perigonium glumaceous.*

1. Juncus.—Style with 3 filiform stigmas. Fruit a capsule, ∞-seeded.
2. Luzula.—Style with 3 filiform stigmas. Fruit a capsule, 3-seeded.

## II. Trigynia.

1. Zygadenus.—Perigonium 6-sepalous. Sepals glandular at the base. Anthers extrorse, 1-celled.
2. Xerophyllum.—Perigonium 6-sepalous. Sepals not glandular at the base. Anthers extrorse.
3. Triglochin.—Ovaries 3-6, separated at the base. Stigmas sessile, plumose.
4. Chorizanthe. Involucre 1-3 flowered, tubular. Fruit an akene.
5. Rumex.—Ovary 1, 1-celled, 1-seeded. Stigma plumose.
6. Pterostegia.—Involucre 2-lobed, 1-flowered. Fruit an akene enveloped by the enlarged involucre.

7. Anemopsis.— Spadix with a several-leaved involucre. Flowers bracteate. Perigonium 0. Ovaries immersed in the rachis, 1-celled, ∞-seeded.
8. Frankenia.—Calyx tubular, costate, 5-cleft. Petals 5, unguiculate. Capsule 1-celled, dehiscent by valves. Ovules few, inserted on parietal placentæ.

### III. Hexagynia—Polygynia.

1. Triglochin.—Stigmas sessile, plumose. Aquatic.
2. Alisma.—Sepals 3. Petals 3. Aquatic.

## HEPTANDRIA.

### I. Monogynia. (Number of parts somewhat inconstant.)

1. Trientalis.—Calyx 7-cleft. Corolla 7-parted. Stamens inserted on the corolla.
2. Æsculus.—Calyx 5-dentate. Petals 4-5, irregular.

### II. Trigynia.

1. Portulaca.—Corolla superior, Calyx 2-sepalous, deciduous. Petals 4-6. Fruit a circumscissile capsule. Succulent.
2. Calandrinia.— Corolla inferior. Calyx 2-sepalous, persistent. Fruit a capsule, 3-valved, ∞-seeded.

### III. Polygynia.
1. Portulaca.—Corolla superior. Calyx 2-sepalous, deciduous. Petals 4-6. Fruit a circumscissile capsule. Succulent.

## OCTANDRIA.
### I. Monogynia.
A. *Floral parts polypetalous, inferior.*
1. Acer.—Calyx 4-5-parted. Corolla 4-5 petals. Fruit a samara. Tree.

B. *Floral parts polypetalous, superior.*
1. Jussiæa.—Calyx tube not prolonged beyond the ovary, lobes of the calyx limb persistent. Petals inserted on the calyx. Stigma capitate. Fruit a septicidal capsule. Aquatic.

   a. *Seeds with a hairy crown.*

1. Zauschneria.—Calyx tube considerably prolonged beyond the ovary; calyx limb petaloid, deciduous. Petals not exceeding the calyx lobes, 2-cleft, erect. Capsule 4-valved, 1-celled, ∞-seeded.
2. Epilobium.—Calyx tube prolonged but little beyond the ovary. Petals inserted on an annular disc on the summit of the calyx tube. Ovary 4-celled. Capsule loculicidal.

b. *Seeds naked.*

1. Gayophytum.—Calyx tube not prolonged beyond the ovary. Anthers versatile. Stamens opposite the petals smaller than alternates. Ovary 2-celled. Style short. Capsule 2-celled, 4-valved.
2. Œnothera.—Calyx tube prolonged beyond the ovary, lobes of the limb reflexed. Anthers versatile. Ovary 4-celled, ovules ∞. Style filiform. Capsule 4-celled, loculicidal.
3. Godetia.—Calyx tube prolonged beyond the ovary, funnel-shaped, lobes of the limb reflexed. Anthers basi-fixed.
4. Clarkia.—Calyx tube prolonged beyond the ovary, lobes of the limb reflexed. Petals unguiculate. Stamens alternately smaller. Anthers basi-fixed.
5. Boisduvalia.—Calyx tube prolonged beyond the ovary, lobes of the limb erect. Petals sessile, 2-lobed. Alternate stamens shorter. Anthers basi-fixed.

C. *Floral parts complete, superior. Corolla gamopetalous.*

1. Vaccinium.—Stamens inserted on the margin of an epigynous disc. Cells of anthers elongated into a tube at the apex. Fruit a ∞-seeded berry. Shrub.

D. *Floral parts incomplete, inferior.*
1. Dirca.--Perigonium campanulate. Limb obliquely truncate. Style subterminal. Fruit a drupe. Shrub.
2. Polygonum.—Perigonium 4-5 sepalous, persistent. Stigmas 2-3. Fruit an akene, enclosed in the persistent perigonium.
3. Acœna.—Calyx tube contracted at the throat, angular, the angles armed with glochidate prickles. Stigma penicillate. Akene enclosed in the calyx-tube.

II. **Digynia and Trigynia.**
1. Polygonum.—Perigonium 4-5 sepalous, persistent. Fruit an akene enclosed in the perigonium.

III. **Tetragynia.**
1. Anemopsis.—Spadix with a several leaved involucre. Perigonium 0. Ovaries immersed in the rachis, 1-celled ∞-ovuled.

### ENNEANDRIA.
1. Oreodaphne.—Anthers with valvular dehiscence. Fruit a drupe resting on an enlarged thalamus. Tree.
2. Eriogonum—.Involucre ∞-flowered, campanulate, 6-dentate. Styles 3. Fruit an akene.

3. Chorizanthe.—Involucre 1-3-flowered, tubular. Styles 3. Fruit an akene.

## DECANDRIA.

**I. Monogynia.**

**A.** *Corolla 4-5 petalous, or petals 0. Never gamopetalous.*

1. Pyrola.—Calyx 5-parted, persistent. Hypogynous disc 0. Anthers dehiscent by 2 pores. Seeds minute.
2. Cercis.—Calyx scarcely dentate. Corolla papilionaceous. Anthers versatile. Petals unguiculate, carina 2-petalous, larger than the upper petals. Legume flat ∞-seeded. Ventral suture winged. Shrub. Tree.
3. Thermopsis.—Calyx cleft to the middle. Corolla papilionaceous. Petals of the carina partly connate, longer than vexillum. Legume linear, compressed.
4. Pickeringia.—Calyx repandly 4-dentate. Corolla papilionaceous. Carina 2-petalous, as long as vexillum. Legume linear, compressed. Shrub.
5. Geranium.—Carpidia 5, verticillate around a columnar axis. Styles connate at the apex. Stamens all fertile.
6. Erodium.—Carpidia 5, verticillate

around a columnar axis. Styles connate at the apex. Alternate stamens sterile.

7. Limnanthes.—Stamens inserted on a perigynous disc. Carpidia 5, each 1-ovulate. Style entire from its base. Aquatic.
8. Jussiæa.—Petals inserted on the calyx. Stigma capitate. Fruit a septicidal capsule. Aquatic.
9. Adenostoma.—Calyx funnel-shaped, the tube 10-costate. Carpidion 1. Fruit an akene enclosed in the calyx tube. Shrub.
10. Acæna.—Calyx tube contracted at the throat, angular, the angles armed with glochidiate prickles. Petals 0. Stigma penicillate. Fruit an akene enclosed in the calyx tube.

B. *Corolla gamopetalous.*

1. Rhododendron.—Calyx 5-parted. Corolla funnel-shaped, rotate. Stamens ascending. Anthers dehiscent by an apical pore. Capsule septicidal. Shrub. Tree.
2. Gaultheria.—Calyx 5-cleft. Corolla urceolate. Fruit a spurious berry, that is, a capsule, 5-celled, ∞-seeded, enclosed in the enlarged and fleshy calyx. Shrub.

3. Arctostaphylos.—Corolla urceolate. Ovary 5-celled, cells 1-ovulate. Fruit baccate. Shrub.
4. Arbutus.—Corolla urceolate. Ovary 5-celled, cells ∞-seeded. Fruit a berry. Tree.
5. Samolus.—Alternate stamens sterile. Ovary 1-celled, ∞-seeded with central placentation.

## II. Digynia.

1. Saxifraga.—Ovary 2-celled, ∞-seeded. Capsule loculicidal.
2. Tellima.—Petals lobed. Ovary 1-celled, ∞-seeded. Styles short. Stigmas capitate. Capsule dehiscent at the apex.
3. Tiarella.—Petals entire. Ovary 1-celled. Styles short. Stigmas simple. Capsule dehiscent to the base; valves unequal.
4. Acæna.—Calyx tube contracted at the throat, angular, the angles armed with glochidiate prickles. Petals 0. Stigmas penicillate. Akenes 2, enclosed in the calyx tube.

## III. Trigynia.

1. Silene.—Calyx gamosepalous, 5-dentate, without bracts at the base. Petals unguiculate. Capsule dehiscent by teeth. Seeds reniform.

2. Alsine.—Sepals 5, unchanged in fruit. Petals entire. Capsule dehiscent by 3 valves. Seeds reniform.
3. Spergularia (Lepigonum). — Sepals 5. Petals entire. Capsule dehiscent by 3 valves. Seeds compressed. Stipules membranaceous.
4. Arenaria. —Sepals 5. Petals entire. Capsule dehiscent by 6 valves. Seeds reniform.
5. Stellaria. — Sepals 5. Petals 2-cleft. Capsule dehiscent by 6 valves.
6. Portulaca.—Corolla superior. Calyx 2-sepalous, deciduous. Petals 4-6. Capsule circumscissile.
7. Calandrinia.—Corolla inferior. Calyx 2-sepalous, persistent. Capsule 3-valved, ∞-seeded.
8. Whipplea.—Ovary 3-septate. Cells 1-ovulate. Capsule septicidal.

## IV. Tetragynia.

1.—Sagina.—Sepals 4. Petals entire. Stigmas alternate with the sepals. Capsule dehiscent by 4–5 valves.

## V. Pentagynia.

1. Sagina.—Sepals 5. Petals entire. Stigmas alternate with the sepals. Capsule dehiscent by 5 valves. Seeds reniform. Stipules 0.

2. Spergula.—Sepals 5. Petals entire. Stigmas alternate with the sepals. Capsule dehiscent by 5 valves. Valves opposite to the sepals. Stipules membranaceous.
3. Cerastium. — Sepals 5. Petals 2-cleft. Capsule dehiscent by 10 teeth.
4. Oxalis.—Stamens monadelphous at their base. Capsule 5-lobed.
5. Cotyledon (Echeveria).—Calyx 5-parted. Petals coherent by their claws. Ovaries 5. Fruit 5-follicles. Succulent.
6. Sedum.—Calyx 5-parted. Petals not coherent by their claws. Ovaries 5. Fruit 5-follicles. Succulent.
7. Portulaca.—Corolla superior. Calyx 2-sepalous, deciduous. Petals 4-6. Capsule circumscissile. Succulent.

## VI. Polygynia.
1. Horkelia.—Calyx companulate, limb 5-parted, augmented by 5 bractlets. ovaries $\infty$, inserted on a conical receptacle. Styles subterminal.

## DODECANDRIA.
### I. Monogynia.
1. Asarum.—Perigonium 3-cleft, superior.
2. Portulaca.—Calyx 2-sepalous, deciduous. Petals 4-6. Capsule circumscissile. Succulent.

3. Calandrinia.—Corolla inferior. Calyx 2-sepalous, persistent. Capsule 3-valved, ∞-seeded. Succulent.
4. Lythrum.—Calyx tubular. inferior, 8-12-dentate. Petals 4—6.
5. Adenostoma.—Calyx funnel-shaped, tube 10-costate. Petals 5. Fruit an akene enclosed in the calyx tube. Shrub.
6. Cercocarpus.—Calyx tube cylindrical. Petals 0. Style terminal, long-exserted. Fruit an akene, linear, terete, caudate by the elongated plumose style. Shrub.

## II. Pentagynia.

1. Nuttallia.—Ovaries 5, each 2-ovulate. Styles subterminal. Fruit several drupes. Shrub.

## ICOSANDRIA.

### I. Monogynia.

A. *Floral parts complete, inferior.*

1. Prunus.—Fruit a fleshy drupe. Endocarp not rugose. Shrubs. Tree.
2. Adenostoma.—Calyx funnel-shaped; tube 10-costate. Petals 5. Akene enclosed in the calyx tube. Shrubs.
3. Cercocarpus.—Calyx tube cylindrical. Petals 0. Style terminal, long-exserted. Akene linear, terete, caudate by the elongated exserted plumose style. Shrub.

4. Portulaca.—Calyx 2-sepalous, deciduous. Capsule circumscissile.
5. Calandrinia.—Calyx 2-sepalous, persistent. Capsule 3-valved, ∞-seeded.
6. Eschscholtzia.— Calyx gamosepalous, circumscissile. Ovary terete. Stigmas 4. Capsule 1-celled, 2-valved, seeds on the margins of the valves.

**II. Di-Tri-Tetra-Pentagynia.**
1. Nuttallia.—Ovaries 5, each 2-ovulate. Style subterminal, subinternal. Fruit several drupes. Shrub.
2. Photinia (Heteromeles).—Ovaries 2, imperfectly united, half superior. Styles 2. Fruit a pomum, 2-celled. Cells 1-seeded. Tree.
3. Amelanchier.—Ovary inferior, 3–5-celled, each cell imperfectly divided. Fruit a pomum, 3–5-seeded. Endocarp bony (drupe). Shrub.
4. Spiræa.—Ovaries several, several seeded. Fruit several follicles. Testa of the seeds membranaceous. Shrub.
5. Neillia.—Ovaries several, several seeded. Fruit several follicles. Testa of the seed crustaceous. Shrub.
6. Horkelia.—Calyx augmented by 5 bractlets. Ovaries on a conical receptacle. Styles subterminal. Fruit akenes.

7. Mesembrianthemum.—Calyx partitions unequal. Petals ∞. Ovary ∞-celled, ∞-seeded. Fruit a capsule. Succulent.
8. Hypericum.—Calyx 4-5-parted. Petals 4-5. Fruit a capsule.
9. Mentzelia.—Ovary inferior. Calyx limb 5-parted, persistent. Capsule 1-celled. Placentæ 3, parietal.

### III. Polygynia.

1. Spiræa.—Ovaries superior, several seeded. Fruit several follicles. Shrub.
2. Rosa.—Calyx 5-parted. Petals 5. Fruit akenes enclosed in the calyx tube forming a spurious berry. Shrub.
3. Calycanthus.—Calyx cup-shaped, limb ∞-parted, ∞-seriate. Petals 0. Fruit ∞ akenes enclosed in the persistent calyx tube. Shrub.
4. Rubus.—Calyx 5-parted. Petals 5. Fruit ∞ drupes on a conical receptacle. Shrubs.
5. Horkelia.—Calyx campanulate, augmented by 5 bractlets. Styles subterminal. Fruit ∞ akenes.
6. Potentilla.—Calyx flattened, augmented by 5 bractlets. Ovaries on a slightly conical receptacle. Styles lateral. Akenes on a herbaceous thalamus.

7. Fragaria.—Calyx flattened, augmented by 5 bractlets. Ovaries on a convex receptacle. Styles lateral. Akenes on a fleshy receptacle.
8. Mesembrianthemum.—Calyx partitions unequal. Petals $\infty$. Ovary $\infty$-celled, $\infty$-seeded. Fruit a capsule. Succulent.

## POLYANDRIA.

### I. Monogynia.

A. *Petals 4.*

1. Dendromecon.—Calyx 2-sepalous. Ovary linear. Stigma sessile, 2-lobed. Capsule 1-celled, 2-valved, seeds on the margin of the valves. Shrub.
2. Eschscholtzia.—Thalamus top-shaped. Calyx gamosepalous, circumscissile. Ovary terete. Stigmas 4. Capsule 1-celled, 2-valved, seeds on the margin of the valves.
3. Meconopsis.—Calyx 2-sepalous, caducous. Anthers laterally dehiscent. Ovary 1-celled. Placentæ more than 2. Style short. Stigma radiate.
4. Argemone.—Calyx 2–3 sepalous. Petals 4. Anthers extrorse. Ovary 1-celled, placentæ more than 2. Style 0. Stigma radiate.

5. Actæa.—Calyx 4-sepalous. Anthers adnate, introrse. Ovary 1. Placenta ventral. Style 0. Berry ∞-seeded.

B. *Petals 5.*

1. Helianthemum.—Calyx double. External sepals 2, sometimes 0; internal 3, contorted in æstivation. Capsule 3-valved.

C. *Petals 6.*

1. Argemone.—Calyx 3-sepalous. Filaments filiform. Stigmas radiate.
2. Platystemon.—Calyx 3-sepalous. Filaments flattened their entire length, ovary at last separating into its component carpidia.
3. Platystigma.—Calyx 3-sepalous. Filaments flattened only at the base, ovary not separating but ripening into a 3-valved capsule.

II. Di-, Tri-, Tetra-, Penta-, Polygnia.

A. *Ovaries ventrally dehiscent. Fruit a follicle.*

    a. *Flowers irregular.*

1. Delphinium. Calyx petaloid. Superior sepal calcarate.

    b. *Flowers regular.*

1. Platystemon.—Sepals 3, caducous. Petals 6. Filaments flattened. Carpidia at first united into an ∞-celled ovary.

2. Pæonia.—Sepals 5, herbaceous, carpidia distinct from the beginning.
3. Aquilegia.—Sepals petaloid. Petals funnel-shaped, calcarate.

B. *Ovaries 1-ovulate, indehiscent. Fruit, akenes.*

    a. *Æstivation of sepals imbricate.*

1. Ranunculus.—Petals unguiculate, claws shorter than the lamina, a nectarium near each claw.
2. Myosurus.—Claws of the petals longer than the lamina.
3. Anemone.—Sepals petaloid. Petals 0. Thalamus convex, conical.
4. Thalictrum. Sepals petaloid. Petals 0. Thalamus flat, discoid.

    b. *Æstivation of sepals valvate.*

1. Clematis.—Sepals 4, petaloid. Petals 0.

**DIDYNAMIA.**

I. **Gymnospermia.**

A. *Stamens and style enclosed. An interrupted ring of hairs in the corolla below the insertion of the stamens.*

1. Marrubium.—Akenes flat at their tops.

B. *Style always exserted. An uninterrupted ring of hairs in the corolla below the insertion of the stamens.*

a. *Calyx 2-labiate. Stamens parallel.*

1. Prunella.—Calyx, closed in fruit.

b. *Calyx 5-dentate. Stamens parallel.*

1. Stachys.—Calyx 5–10-nerved. Lobes of the lower lip of the corolla obtuse. Stamens ascending.
2. Sphacele.—Calyx reticulately veined. Lobes of the corolla 5, all obtuse. Stamens ascending. Shrub.

c. *No ring of hairs in the corolla.*

1. Scutellaria.—Upper lip of the corolla concave. Stamens close together, parallel. Calyx 2-labiate, lips entire.
2. Mentha.— Stamens distant and divergent, but straight. Corolla funnel-shaped, 4-cleft, superior lobe emarginate.
3. Micromeria.—Calyx teeth nearly equal. Stamens ascending.
4. Monardella.—Stamens straight, divergent, exserted. Cells of the anthers at length also divergent. Corolla lobes narrow.
5. Pycnanthemum.—Stamens straight, distant, divergent. Cells of the anthers parallel to the last.
6. Pogogyne.—Stamens ascending, convergent in pairs. Style villous. Calyx deeply 5-cleft.

7. Acanthomintha.—Lower pair of stamens ascending. Cells of the anthers divaricate. Calyx 2-labiate, teeth spinose.
8. Lophanthus.—Upper pair of stamens longer and declined, lower pair shorter and ascending. Anther cells parallel. Calyx 15-nerved.
9. Trichostemma.—Stamens long-exserted, lower pair longer. Tube of corolla slender; limb 5-cleft, lobes oblong, declined.

**II. Angiospermia.**

A. *Base of the anther cells mucronate.*

    a. *Ovary 1-celled $\infty$-ovulate, placentæ parietal.*

1. Anoplanthus.—Flowers without bractlets. Parasitic. Chlorophyll 0.
2. Aphyllon.—Flowers with bractlets. Parasitic. Chlorophyll 0.

    b. *Ovary 2-celled, $\infty$-ovulate. Fruit 1-$\infty$-seeded.*

1. Pedicularis.—Calyx 5-dentate, superior lip of corolla laterally compressed.

B. *Base of anther cells not mucronate.*

    a. *Ovary 1-celled. Placenta $\infty$-seeded, central.*

1. Limosella.—Calyx 5-dentate. Corolla subregular.

b. *Ovary 2-celled.*

1. Cordylanthus.—Calyx spathaceous. Cells of anthers distant and of different shape and insertion. Lips of corolla short, upper lip laterally compressed.
2. Orthocarpus.—Calyx spathaceous, cleft vertically. Cells of anthers distant and of different shape and insertion. Corolla lips personate, but upper lip the smaller.
3. Castilleia.—Calyx spathaceous, cleft vertically. Cells of anthers distant and of different shape and insertion. Corolla ringent, the upper lip the larger.
4. Mimulus.—Stigma dilated, petaloid. Calyx angular, angles carinate. Placentæ of the capsule remain united and only separate at last near the apex. Valves of the capsule membranaceous.
5. Mimetanthe.—Stigma dilated, petaloid. Calyx campanulate, not angulate. Capsule dehiscent only by the dorsal suture.
6. Diplacus. — Stigma dilated, petaloid. Calyx angular; angles carinate. Placentæ of the capsule meeting, but in dehiscence separating their whole length. Valves of the capsule coriaceous. Tube of the corolla funnel-shaped.

7. Eunanus. — Stigma dilated, petaloid. Calyx angular, angles carinate. Placentæ of the capsule separate in dehiscence for their whole length. Tube of corolla slender, filiform.

8. Collinsia.—Stigma small. Corolla personate, lower lip 3-lobed, middle lobe laterally compressed, carinate, hiding the stamens.

9. Scrophularia.—Stigma small. Corolla short, globular. Middle lobe of lower lip reflexed.

10. Pentstemon. — Stigma small. Corolla ringent. One sterile stamen besides the didynamous ones.

11. Antirrhinum.—Corolla personate, saccate at the base. Capsule opening by pores.

12. Linaria.—Corolla personate, calcarate. Capsule dehiscent by valves.

13. Lippia.—Calyx 2-cleft. Cells of ovary 1-seeded. Fruit drupaceous, separating into 2 nutlets.

c. *Ovary 4-celled.*

1. Verbena. — Cells of ovary 1-ovulate. Fruit drupaceous, splitting into 4 parts.

## TETRADYNAMIA.

**I. Siliculosa.**

A. *Fruit a 2-articulate lomentum.*
  1. Cakile.—Both articles of the lomentum anceps.

B. *Fruit a schizocarp.*
  1. Senebiera.—Silicula 2-seeded, margin not winged.

C. *Fruit an indehiscent silicula.*
  1. Thysanocarpus.—Silicula 1-seeded, margin winged.

D. *Silicula regularly dehiscent, dorsally compressed, septum narrow.*
  1. Lepidium.—Cells 1-seeded.
  2. Capsella.—Cells ∞-seeded.

E. *Silicula regularly dehiscent, laterally compressed, septum broad.*
  1. Alyssum.—Cells 1-4-ovulate.

**II. Siliquosa.**

A. *Siliqua indehiscent.*
  1. Raphanus.—Siliqua moniliform.

B. *Siliqua regularly dehiscent.*
  1. Cardamine.—Siliqua laterally compressed. Valves linear, without nerves. Ovules in each cell ∞, 1-seriate.

2. Nasturtium. — Siliqua linear. Valves without nerves. Ovules in each cell ∞, 2-seriate.
3. Cheiranthus. — Siliqua laterally compressed. Septum broad. Valves 1-nerved. Ovules in each cell 1-seriate. Lateral sepals saccate. Petals unguiculate.
4. Caulanthus.—Siliqua terete, valves 1-nerved. Ovules in each cell 1-seriate Lateral sepals saccate. Petals undulate.
5. Streptanthus. — Siliqua laterally compressed. Valves 1-nerved. Ovules in each cell 1-seriate. All sepals equal. Petals undulate.
6. Barbarea.—Siliqua 4-angular. Valves concave, 1-nerved. Ovules in each cell 1-seriate.
7. Erysimum.—Siliqua 4-angular. Valves carinate, 1-nerved. Ovules in each cell 1-seriate.
8. Tropidocarpum.—Siliqua dorsally compressed. Valves 1-nerved, carinate. Septum very narrow. Ovules in each cell 2-seriate.
9. Arabis.—Siliqua laterally compressed. Valves 1-nerved. Ovules in each cell 1-seriate. Petals not undulate.

10. Sisymbrium.—Siliqua linear. Valves 3-nerved, concave. Ovules in each cell 1-seriate.
11. Brassica.— Siliqua terete. Valves 3-5-nerved, concave. Ovules in each cell 1-seriate.

**MONADELPHIA.**

I. **Triandria.**
 1. Sisyrinchium.

II. **Pentandria.**
 1. Erodium. — Stamens alternating with staminodia.
 2. Downingia.—Corolla gamopetalous, irregular.

III. **Octandria.**
 1. Polygala.

IV. **Decandria.**
 1. Erodium. — Alternate stamens sterile. Fruit 5 akenes with persistent styles.
 2. Geranium.—Stamens all fertile. Fruit 5 akenes with persistent styles.
 3. Oxalis.—Styles 5. Fruit a 5-lobed capsule.
 4. Lupinus.—Corolla papilionaceous.
 5. Amorpha.—Corolla represented by a single petal. Shrub.

## V. Polyandria.

1. Lavatera. — Calyx double. External calyx 3-6-cleft. Shrub.
2. Malva.—Calyx double. External calyx 3-sepalous.
3. Sidalcea.—External calyx 0. Stigmas not capitate.
4. Sphæralcea.—Stigmas capitate. Cells of capsule 2-ovulate. Shrub.
5. Sida.—Stigmas capitate. Cells of capsule 1-ovulate.

## DIADELPHIA.

### I. Hexandria.

1. Dicentra.—Flowers 2-calcarate.

### II. Octandria.

1. Polygala.—Flowers irregular. Ovary 2-celled. Cells 1-ovulate.

### III. Decandria.

A. *Sub-monadelphous*.

1. Lupinus.—Anthers alternately oblong and reniform.
2. Amorpha.—Corolla represented by a single petal. Shrub.
3. Psoralea.—Wings of the corolla united to the carina. Ovary sessile, 1-ovulate. Legume indehiscent, included in the calyx.

B. *Genuinely diadelphous.*

    a. *Style not pilose.*

1. Hosackia.—Petals unguiculate. Carina rostrate.
2. Trifolium.—Petals persistent, adnate to the stamineal tube.
3. Psoralea.—Wings of the corolla adnate to the carina. Ovary sessile, 1-ovulate. Legume indehiscent, included in the calyx.
4. Glycyrrhiza.—Petals not adnate to the stamineal tube. Carina 2-petalous.
5. Astragalus.—Petals not adnate to the stamineal tube. Carina blunt. Legume imperfectly 2-celled.
6. Melilotus.—Petals not adnate to the stamineal tube. Ovary straight. Legume globose.
7. Medicago.—Petals not adnate to the stamineal tube. Ovary crescent-shaped. Legume spiral.

    b. *Style pilose.*

1. Lathyrus.—Style pilose only on the ventral side which is flattened.
2. Vicia.—Style pilose on the dorsal side, or all around the apex. Style not spiral.

## POLYADELPHIA.

1. Hypericum. — Ovary superior. Calyx 4-5-parted. Petals 4-5. Fruit capsular.
2. Mentzelia.—Ovary inferior. Calyx limb 5-parted, persistent. Capsule 1-celled. Placentæ 3, parietal.

## SYNGENESIA.

I. Æqualis.

A. *Flowers ligulate.*

    a. *Pappus plumose.*

    aa. *Receptacle paleaceous.*

1. Hypochæris.—Paleæ of the receptacle deciduous.

    bb. *Receptacle not paleaceous.*

1. Microseris.—Pappus paleaceous at base. Akenes more than 5-costate, truncate at apex. Scapigerous.
2. Stephanomeria. — Pappus setaceous, somewhat plumose. Akenes 5-costate, truncate at apex and base. Caulescent.
3. Rafinesquia.—Bristles of pappus arachnoid. Akenes rostrate.
4. Malacothrix.—Pappus 1-seriate. Bristles scabrous, silky, deciduous in a ring. Apex of akenes developed into a crown.

5. Troximon.—Pappus capillary. Bristles deciduous singly, not in a ring. Apex of akenes rostrate; costæ smooth.
6. Taraxacum.—Pappus capillary. Bristles persistent. Apex of akenes elongated, rostrate; costæ roughened.
7. Sonchus.—Akenes flattened, not rostrate. Pappus silky, white.
8. Hieracium.—Akenes terete, not rostrate, Pappus brittle, discolored.

B. *Flowers tubular.*

   a. *Receptale naked.*

   aa. *Branches of the style club-shaped.*

1. Adenostyles. — Involucre 1-seriate. Branches of the style filiform. Pappus copious, silky, white.
2. Bulbostylis.—Involucre imbricate, 2-3 seriate. Branches of the style filiform. Pappus copious, silky.

   bb. *Branches of the style flat and tipped with an appendix.*

   aaa. *Parts of pappus $\infty$.*

1. Bigelovia.—Involucre imbricate. Pappus 1-seriate, capillary, brownish at maturity.

2. Aplopappus.—Involucre imbricate. Pappus 1-seriate. Bristles rigid.
3. Chrysopsis.—Involucre imbricate. Pappus 2-seriate, internal capillary, external scaly.
4. Lessingia.—Ray-flowers enlarged, but not ligulate.
5. Baccharis.—Diœcious. Shrub.

bbb. *Parts of pappus definite in number.*

1. Pentachæta. — Pappus of 5-scabrous bristles, shorter than the corolla, sometimes depauperate.
2. Grindelia.—Pappus aristate.

cc. *Branches of style not flat, but tapering, sometimes with an appendix.*

aaa. *Scales of the involucre herbaceous.*

1. Chænactis.—Pappus paleaceous.
2. Senecio.—Pappus capillary.

bbb. *Scales of the involucre scarious.*

1. Artemisia.—Receptacle slightly convex. Pappus 0. Akenes obovate with small epigynous disc.
2. Tanacetum.—Receptacle slightly convex. Pappus 0. Akenes costate with large epigynous disc.

3. Matricaria.—Receptacle cylindrical, hollow.
4. Baccharis.—Diœcious.

   b. *Receptacle paleaceous or setaceous.*

   aa. *Branches of style connected near the apex.*

1. Cnicus. — Receptacle densely bristly. Bristles of the pappus plumose, connected at the base into a deciduous ring. Filaments of stamens distinct.
2. Silybum.— Receptacle densely bristly. Bristles of pappus not plumose, 1-seriate.
3. Carthamus.—Pappus 0.
4. Centaurea.—Receptacle densely bristly. Bristles of pappus not connected at their base into a ring.

   bb. *Branches of style entirely distinct.*

1. Bidens.— Involucre 2-seriate, external spreading, internal erect. Pappus of 2-4 awns.
2. Madia.— Pappus 0. Receptacle sometimes only paleaceous near the ray and naked at the disc.

## II. Superflua.
  A. *Receptacle naked.*

    a. *Anthers obtuse at their base, at least not calcarate.*

**aa.** *Branches of the style flat and tipped with an appendix.*

**aaa.** *Parts of pappus* ∞.

1. Bigelovia.—Involucre imbricate; scales dry. Pappus 1-seriate, capillary, at maturity brownish. Ray flowers yellow.
2. Aplopappus.—Involucre imbricate, scales foliaceous. Pappus 1-seriate, bristles rigid. Ray flowers yellow.
3. Chrysopsis.—Involucre imbricate. Pappus 2-seriate, internal capillary, external scaly. Ray flowers yellow.
4. Lessingia. – Ray flowers enlarged but not ligulate.
5. Solidago.—Involucre imbricate, scales foliaceous. Pappus silky. Ray flowers yellow.
6. Aster.—Involucre imbricate, scales foliaceous. Ray flowers not yellow, 1-seriate.
7. Erigeron.—Involucre imbricate, scales foliaceous. Ray flowers not yellow, ∞-seriate.

**bbb.** *Parts of pappus definite in number.*

1. Pentachæta.—Scales of involucre not herbaceous. Pappus of 5 scabrous bristles shorter than the corolla, sometimes depauperate. Rays yellow.

2. Grindelia.—Scales of the involucre herbaceous. Pappus aristate. Rays yellow.
3. Gutierrezia.—Scales of the involucre herbaceous. Pappus short, paleaceous. Rays yellow.

bb. *Branches of the style not flat but tapering, sometimes with an appendix.*

1. Coinogyne.—Involucre regularly imbricate. Scales broad.
2. Burriellia.—Involucre consisting of 3-5 scales.
3. Bæria.—Scales of the involucre in a single series. Receptacle conical, minutely muricate. Akenes linear.
4. Eriophyllum (Bahia).—Scales of the involucre not embracing any akenes. Receptacle convex (not conical). Akene linear, with large terminal areola.
5. Monolopia.—Scales of involucre united at base. Receptacle conical. Akene oblong, with small terminal areola.
6. Lasthenia.—Scales of involucre united into a dentate cup. Receptacle conical. Pappus 0.
7. Rigiopappus.—Scales of involucre linear. Receptacle flat. Pappus 4-5 aristate scales.

8. Helenium.—Scales of the involucre reflexed. Pappus of hyaline scales. Akenes turbinate, costate.
9. Senecio.—Scales of the involucre 1-seriate, sometimes calyculate. Pappus capillary, copious.
10. Arnica.—Scales of involucre lanceolate, linear, all equal. Pappus capillary, 1-seriate.

    b. *Anthers calcarate.*

1. Pluchea.—Scales of involucre imbricate, herbaceous.
2. Micropus.—Scales of involucre scarious, ♀ flowers enclosed in marginal scales, which are laterally compressed and conduplicate. Pappus 0.
3. Stylocline.—Scales of involucre scarious. Receptacle cylindrical. ♀ flowers protected by carinate scales.
4. Filago.—Scales of involucre scarious. Scales subtending ♀ flowers not carinate.
5. Anaphalis.—Scales of involucre scarious. ♀ flowers filiform. Style of ☿ flowers nearly blunt.
6. Gnaphalium.—Scales of involucre scarious. ♀ flowers filiform. Style of ☿ flowers 2-cleft.

B. *Receptacle paleaceous.*

    a. *Anthers obtuse or sagittate at their base, but never calcarate.*

    aa. *Involucre 1-seriate or nearly so.*

1. Iva.— ♀ flowers reduced to short tubes embracing the base of the style.  ☿ flowers: anthers scarcely connate.
2. Balsamorrhiza.—Akenes of the disc 4-angular. Pappus 0.
3. Wyethia.—Akenes 4-angular or laterally compressed. Pappus paleaceous, crown-shaped.
4. Pugiopappus.—Akenes villous. Pappus 2-aristate.
5. Leptosyne.—Akenes naked, margined. Pappus almost 0.
6. Madia.—Involucre in a simple row, each scale wrapped about a laterally compressed akene. Pappus 0.
7. Holozonia.—Involucre in a simple row, each scale completely embracing an akene. Pappus of the ray akenes hyaline, of the disc generally 0.
8. Hemizonia.—Involucre 1-seriate, each scale embracing with its base half of a ray akene. Pappus of the ray 0.

9. Lagophylla.—Ray flowers 5 in number, cuneiform, 3-lobed. Involucre 5 scales, each entirely embracing a ray akene. Receptacle flat.
10. Layia.—Capitulum ∞-flowered. Involucre 1-seriate. Scales acuminate, each entirely enclosing a ray akene. Pappus of the ray 0.
11. Achyrachæna.—Ray flowers short and hidden by the scales of the 1-seriate involucre. Receptacle naked in the centre, paleaceous toward the margin. Pappus paleaceous. Paleæ ∞-seriate, silvery.

   bb. *Involucre imbricate.*

1. Achillea.—Involucre of few series of scales. Rays short, oval.
2. Anthemis.—Involucre ∞-seriate. Rays long, ligulate.

   b. *Anthers calcarate.*

1. Psilocarphus.—Scales of the involucre few. ♀ flowers each wrapped in a palea. Pappus 0.
2. Stylocline.—Scales of the involucre few. ♀ flowers each bracteate by a carinate palea, but not wrapped in it. Pappus a few caducous bristles.

3. Evax.—Scales of the involucre 1-2 seriate. Pappus 0.
4. Filago.—Scales of the involucre imbricate. Pappus pilose.

### III. Frustranea.
1. Corethrogyne.—Receptacle naked.
2. Centaurea.—Receptacle bristly.
3. Bidens.—Paleæ of the receptacle deciduous.
4. Helianthus.—Paleæ of the receptacle persistent. Pappus paleaceous, caducous.
5. Helianthella. Paleæ of the receptacle persistent. Pappus 2 persistent awns, besides the caducous paleæ.

### IV. Necessaria.
A. *Receptacle naked.*

    a. *Anthers obtuse at the base, at least not calcarate.*

    aa. *Branches of the style of the ♀ flowers flat and tipped with an appendix.*

1 Erigeron.

    bb. *Branches of the style not flat, but tapering, sometimes with an appendix.*

1. Hemizonia.—Involucre 1-seriate, each scale with its base half embracing a ray akene. Pappus of ray 0.

2. **Holozonia.** — Involucre 1-seriate, each scale completely embracing a ray akene. Pappus of the ray hyaline.
3. **Lagophylla.** — Ray flowers 5, cuneiform, 3-lobed. Involucre of 5 scales, each completely enclosing a ray akene. Receptacle flat. Paleæ 5, between disc and ray. Pappus 0.
4. **Blennosperma.** — Scales of the involucre 1-seriate, united at the base, not embracing the akene. Pappus 0.
5. **Artemisia.** — Involucre imbricate, scarious. Flowers of the ray slender, dentate.
6. **Soliva.** — Flowers of the ray ∞-seriate, without corolla. Pappus 0.
7. **Nardosmia.** — Pappus pilose.

B. *Receptacle villous.*

1. **Artemisia.**

C. *Receptacle paleaceous.*

    a. *Anthers obtuse or sagittate at their base, but not calcarate.*

1. **Lagophylla.** — Ray flowers 5, cuneiform, 3-lobed. Involucre of 5 scales, each completely enclosing a ray akene. Receptacle flat. Paleæ 5, between disc and ray. Pappus 0.
2. **Holozonia.** — Involucre 1-seriate, each scale embracing completely a ray akene. Pappus of the ray hyaline.

3. Hemizonia.—Involucre 1-seriate, each scale with its base half embracing a ray akene. Pappus of ray akene 0.
4. Madia.—Involucre 1-seriate, each scale completely enwrapping a laterally compressed ray akene. Pappus 0.

 b. *Anthers almost distinct.*

1. Iva.

 c. *Anthers calcarate.*

1. Evax.—Scales of involucre 1—2-seriate. Pappus 0.
2. Stylocline.—Scales of involucre very few. Flowers of the ray bracteate by a carinate palea, but not enwrapped by it. Pappus a few caducous bristles.
3. Psilocarphus.—Scales of the involucre few. Each ray flower enwrapped in a palea. Pappus 0.

## V. Monogamia.

1. Downingia.

## GYNANDRIA.

### I. Monandria.

 a. *Labellum calcarate.*

1. Habenaria.

 b. *Labellum not calcarate.*

1. Epipactis.—Labellum geniculate.

2. Corallorhiza. Labellum not geniculate, adnate to the column. Aphyllous and and without chlorophyll.
3. Spiranthes.—Labellum not geniculate, embracing the column with its base. Perigonium oblique.

**II. Diandria.**
1. Cypripedium.

**III. Hexandria.**
1. Aristolochia.

## MONŒCIA.

**I. Monandria.**
1. Euphorbia. — Involucre campanulate, sometimes 2-phyllous. Flowers umbellate, androgynous. ♂ flowers without perigonium, stipitate, bracteate, ♀ flowers single, central. Ovary 3-celled, styles 3, each 2-cleft.
2. Zostera.—Spathe elongated into a lamina. Spadix androgynous. Ovary 1-celled, 1-ovulate. Fruit a nutlet. Aquatic.
3. Lilæa.—Flowers spicate. ♂, ♀, and androgynous spikes present, besides solitary axillary ♀ flowers. Aquatic.
4 Najas. — Flowers axillary, solitary. ♂ flowers included in a spathe ♀ flowers naked. Ovary 1, 1-ovulate. Fruit drupaceous. Aquatic.

5. Zannichellia.—Flowers axillary. ♂ flower naked, ♀ flower with campanulate perigonium. Ovaries 4, each 1-ovulate. Fruit nutlets. Aquatic.
6. Callitriche.—Flowers axillary, 2-bracteolate. Ovary 4-celled; cells 1-ovulate. Fruit a 4-coccous schizocarp. Aquatic.

## II. Diandria.

1. Pinus.—Anthers sessile on the bracts of the ♂ amentum. Leaves fasciculate in a sheath. Coniferous.
2. Pseudotsuga.—Anthers sessile on the bracts of the ♂ amentum. Leaves distichous. Branchlets smooth, the leaf scars not being prominent. Coniferous.
3. Fraxinus.—Flowers paniculate. Calyx obsolete, 4-cleft. Fruit a 1-seeded samara.

## III. Triandria.

1. Typha.—♂ spike and ♀ spike cylindrical. ♂ spike above ♀ spike. Paleæ 0.
2. Sparganium.—♂ spike and ♀ glomerulate. Flowers separated by a paleaceous perigonium. Aquatic.
3. Carex.—Glumaceous. Spikes unisexual or androgynous. Spikelets 1-flowered. ♂ spikelet an external palea; ♀ spikelet with 2 paleæ, the interior one transformed into a utricle including the 1-seeded ovary.

4. Phoradendron.—Flowers immersed in the rachis of articulate spikes. Perigonium 3-lobed. Ovary inferior. Fruit a berry. Parasitic.
5. Megarrhiza.—Flowers complete. ♂ flowers in racemes, ♀ flower solitary in the axilla of the ♂ raceme.

## IV. Tetrandria.

A. *Flowers complete.*

1. Ptelea.—Calyx 4-parted. Petals 4. Stamens alternate with the petals. Fruit a 2-seeded samara. Shrub. Tree.

B. *Flowers perigoniate in ♂ and ♀.*

1. Urtica.— ♀ perigonium 2-sepalous.
2. Hesperocnide.— ♀ perigonium gamosepalous.

C. *Perigonium 0, or only present in one sex.*

1. Cupressus.— ♂ flowers: anthers attached to a peltate bract. Cone globose. Seeds angulate, narrowly winged. Leaves opposite. Coniferous.
2. Libocedrus.— ♂ flowers: anthers attached to a peltate bract. Cone oblong. Wings of the seed unequal. Leaves opposite. Coniferous.

3. Sequoia.—Anthers of the ♂ flowers attached to a peltate bract. Cone ovate. Leaves alternate. Coniferous.
4. Alnus.— ♂ flowers with 3-4 cleft perigonium. ♀ flowers: perigonium 0.

## V. Pentandria—Polyandria.

A. *Flowers complete, not amentaceous.*

1. Myriophyllum.— ♂ flower: stamens 8. ♀ flower: ovary inferior, 4-celled, 4-ovulate. Aquatic.
2. Ptelea.—Calyx 5-parted. Petals 5. Stamens alternating with petals. Style 1. Fruit a 2-seeded samara. Shrub.
3. Rhus.—Calyx 5-parted, persistent. Petals 5. Stamens 5, alternating with the petals. Ovary 1-celled, 1-ovulate. Styles 3. Fruit a drupe.
4. Æsculus.—Calyx irregular, gamosepalous. Petals 5, unguiculate. Filaments ascending. Ovary 3-celled. Style 1. Cells 2-ovulate. Fruit a 1-seeded capsule. Shrub. Tree.
5. Acer.—Stamens 8. Ovary 2-celled, 2-lobed. Fruit a 2-seeded samara. Tree.

B. *Flowers complete. ♂ flowers amentaceous.*

1. Juglans.— ♂ perigonium, 2-6 parted. ♀ flowers with superior 4-dentate calyx. Petals 4. Tree.

C. *Flowers incomplete not amentaceous.*

1. Amarantus.— Perigonium 3-5-sepalous. Fruit a circumscissile capsule.
2. Atriplex.— ☂ perigonium 5-sepalous. ♀ perigonium 2-parted. Stigmas 2. Fruit a compressed utricle.
3. Ceratophyllum.— ☂ perigonium ∞-sepalous. Anthers sessile. Ovary of the ♀ flower 1-celled, 1-ovulate. Fruit a nutlet with persistent style. Aquatic.
4. Eremocarpus.— ☂ perigonium 5-6 parted. ♀ perigonium 0. Ovary 1-celled, 1-ovulate. Fruit a 2-valved capsule.
5. Xanthium.— ☂ inflorescence: involucre ∞-phyllous, ∞-flowered. Perigonium 5-dentate. ♀ inflorescence: involucre gamophyllous, 2-flowered. Fruit a pseudocarp enclosed in the indurated involucre.
6. Ambrosia.— ☂ inflorescence: involucre gamophyllus, ∞-flowered. ♀ inflorescence: involucre gamophyllus, 1-flowered. Spines of the involucre 1-seriate.
7. Franseria.— ☂ inflorescence: involucre gamophyllous, ∞-flowered. ♀ inflorescence: involucre gamoyhyllous, 1-4-flowered. Spines of the involucre ∞-seriate.

8. Iva.—Involucre gamophyllus. Head androgynous. ♂ flowers funnel-shaped, central; ♀ flowers tubular, peripheral.
9. Platanus.—Flowers without perigonium, densely capitulate on a globose receptacle. Tree.

D. *Flowers incomplete; ♂ flowers amentaceous. Stamens inserted on a perigonium.*

1. Castanopsis.—Involucre of ♀ flowers 1-3-flowered, afterward enclosing the nut in the shape of a prickly burr.
2. Quercus.—Involucre of ♀ flowers 1-flowered, surrounding the base of the nut in the shape of a cup. Trees. Shrubs.

E. *Flowers incomplete. ♂ flowers or both ♀ and ♂, amentaceous. Stamens inserted on bract.*

1. Corylus.—Fruit a nut, inclosed in a 2-cleft, laciniate involucre.
2. Myrica.— ♀ flowers amentaceous. Fruit, drupaceous nutlets. Involucre 0. Tree.

### DIŒCIA.

I. Diandria.
1. Salix. — Amentaceous. Perigonium 0. Fruit an ∞-seeded capsule. Shrub.
2. Fraxinus.—Flowers paniculate. Calyx obsolete, 4-cleft. Fruit a 1-seeded samara. Tree.

## II. Triandria.

1. Salix.—Amentaceous. Perigonium 0. Shrub. Tree.
2. Phoradendron.—Flowers immersed in the rachis of an articulate spike. Perigonium 3-lobed. Anthers 2-celled dehiscent by 2 pores. Berry globose, 1-seeded.
3. Arceuthobium.— ♂ perigonium 3-lobed. ♀ 2-dentate. Anthers 1-celled, dehiscent by a slit. Berry compressed, 1-seeded.
4. Atriplex.— ♂ perigonium 3-sepalous. ♀ 2-parted. Fruit a utricle.
5. Amarantus.—Perigonium 3-parted. Fruit a capsule with circumscissile dehiscence.
6. Brizopyrum (Distichlis).—Glumaceous. Inferior glume $\infty$-nerved.
7. Poa.—Glumaceous. Inferior glume 1-nerved.

## III. Tetrandria.

1. Urtica. — ♂ flowers 4-sepalous. ♀ flowers 2-sepalous. Stigma penicillate.
2. Negundo.—Perigonium depauperate, 4-cleft. Ovary 2-lobed, 2-celled. Fruit a 2-seeded samara. Tree.
3. Ptelea.—Calyx 4-5 sepalous. Petals 4. Fruit a 2-seeded samara. Tree.

4. Garrya.—Flowers amentaceous. Bracts decussate, connate in pairs. ♂ flowers 4-parted. Perigonium of ♀ flowers rudimentary. Styles 2. Fruit a berry. Tree.

## IV. Pentandria.

1. Rhus.—Calyx 5-parted, persistent. Petals 5. Stamens alternating with the petals. Ovary 1-celled, 1-seeded. Styles 3. Fruit a drupe. Tree. Shrub.
2. Ptelea.—Calyx 5-parted, petals 5. Stamens alternating with the petals. Style 1. Fruit a 2-seeded samara. Tree.
3. Rhamnus.—Calyx 4-cleft. Petals 0. Stamens alternating with the partitions of the calyx. Fruit 2–4-celled, 2–4-seeded. Shrub.
4. Vitis.—Calyx obsolete. Petals inserted on a disc, coherent at their apex, and deciduous at their base. Fruit baccate.
5. Negundo.—Perigonium depauperate, 5-cleft. Ovary 2-lobed, 2-celled. Fruit a 2-seeded samara. Tree.
6. Juniperus. — ♂ flowers amentaceous. Stamens adnate to the base of a scale. ♀ flowers amentaceous. Ovules naked, 2 adnate to the base of each bract. Fruit a galbulus.

7. Atriplex.— ♂ perigonium 5-sepalous. ♀ perigonium 2-parted. Fruit a utricle.
8. Amarantus.— Perigonium 5-sepalous. Fruit a capsule with circumscissile dehiscence.

## V. Hexandria.
1. Rumex.— Ovary 1-celled, 1-seeded. Styles 3. Stigma plumose.

## VI. Octandria.
1. Populus.—Amentaceous. ♀ flowers cup-shaped. Fruit an ∞-seeded capsule.

## VII. Dodecandria,
1. Nuttallia.—Flowers complete. Ovaries 5. Fruit 1–5 drupes. Shrub.
2. Hendecandra (Croton).— ♂ perigonium 5-cleft. ♀ perigonium 3-cleft. Ovary 3-celled. Fruit a 3-coccous capsule.

## VIII. Icosandria.
1. Nuttallia.—Flowers complete. Shrub.
2. Torreya.— ♂ flowers amentaceous. ♀ flowers single, ovule naked, immersed in an urceolate arillus, in fruit simulating a drupe. Coniferous.

## IX. Polyandria.
1. Thalictrum.

## X. Syngenesia.
1. Baccharis. Shrub.

# SYNOPSIS OF GENERA AND SPECIES.

*Vascular Plants.*

### Class 1. PHANEROGAMÆ.
#### Sub-Class 1. ANGIOSPERMÆ.
### Division 1. DICOTYLEDONES.
### Series 1. TETRACYCLICÆ.

Plants with a tendency to arrangement of the floral parts in 4 well-defined circles, and with a well established numeric (quinary) law.

#### Sub-Series 1. GAMOPETALÆ.
(*Petals Consolidated.*)

Section ANISOCARPÆ. Number of carpidia less than 5.

#### Sub-Section 1. EPIGYNÆ.

ORDER 1—SYNANDRÆ. Stipules 0. Filaments flattened.

#### Family 1. COMPOSITÆ.

Inflorescence capitulate. Calyx changed into a pappus. Anthers syngenetic. Ovary 1-celled, 1-seeded; fruit an akene.

Sub-Family 1. *Tubuliflorœ.*
Flowers, at least of the disc, tubular.

Tribe 1. ASTEROIDEÆ.

Anthers without tails. Style branches of disc-flowers, flat, with an appendix. Leaves alternate.

a. *Parts of the pappus of definite number.*

### 1. Gutierrezia Lag.

Superflua. Heads few-flowered. Ray-flowers about 3, ligulate; disc-flowers about 5. Scales of the involucre coriaceous, imbricate, their tips green, reflexed. Akenes terete. Pappus of about 9, paleaceous scales.—♃. ♄. Flowers yellow.

1. G. CALIFORNICA. Torr. & Gray. Suffrutescent; leaves linear.—Dry river beds.—Sonoma. Livermore. Summer.

### 2. Grindelia Willd. GUM-PLANT.

Superflua. Æqualis. Head ∞-flowered. Ray-flowers ligulate, 1-seriate. Scales of the hemisphærical involucre in several series, their green tips squarrose. Akenes compressed. Pappus aristate.

Resinous herbs; flower buds before opening, bearing a drop of milky-looking resin.—♃. ♄. Flowers yellow.

1. G. HIRSUTULA. Hook. Hirsutely pubescent.—Dry hillsides. Common. Summer.

An infusion of the herb has been recommended against asthma by the Californians of Spanish descent.

2. G. GLUTINOSA Dunal. Glabrous; scales of the involucre with short tips; awns of the pappus 5 or more.—Livermore. Summer.

3. G. CUNEIFOLIA Nutt. Glabrous; scales of the involucre squarrose from the base; awns of the pappus 5 or more; leaves clasping, obtuse.—Salt marshes at Alvarado. Summer.

4. G. ROBUSTA Nutt. Glabrous; scales of involucre squarrose from the base; awns of the pappus less than 5; leaves clasping, acute. Salt marshes. Common. Summer. Has been recommended against the eczema caused by Rhus—(Poison oak.)

### 3. Pentachæta Nutt.

Superflua. Æqualis. Ray-flowers variable, or altogether wanting. Scales of the involucre lanceolate, scarious on the margin and tip, loosely imbricate in about two series. Receptacle convex. Pappus about 5 scabrous bristles shorter than the disc-corolla, sometimes depauperate.—☉. Small vernal herbs.

1. P. BELLIDIFLORA Greene. Peduncles glabrous; ray-flowers white.—Corte Madera.

2. P. EXILIS Gray. Stem erect; peduncles villous; ray-flowers 0; disc-flowers purple. Marin County.

3. P. ALSINOIDES Greene. Stem diffuse, much branched; capitula almost sessile in the axillæ of branches; ray-flowers 0; disc-flowers purple.—Marin County.

b. *Parts of the pappus* ∞.

#### 4. Lessingia Cham.

Æqualis. Superflua. Ray-flowers with palmate limb. Receptacle flat, naked. Akenes compressed, silky-villous; pappus 1-seriate; bristles ∞, scabrous, rigid.—⊙

1. L. GERMANORUM Cham. Limb of ray-flowers unequally lobed; stems spreading on the ground; lower leaves spathulate, pinnatifid; flowers yellow.—Sandhills. San Francisco. Summer.

2. L. RAMULOSA Gray. Limb of the ray-flowers unequally lobed; stem erect, diffusely branched; leaves not pinnatifid; flowers purple.—Marin County. Summer.

3. L. LEPTOCLADA Gray. Limb of ray-flowers equally lobed; stem erect; branches filiform, terminated by the capitula; flowers purple, white.—Nicasio. Summer.

### 5. Chrysopsis Nutt. GOLDEN ASTER.

Superflua. Æqualis. Head $\infty$-flowered; ray-flowers ligulate. Akenes compressed; pappus 2-seriate; outer row short paleaceous scales, inner row long scabrous bristles.—♃. Flowers yellow.

1. C. SESSILIFLORA Nutt. Capitula radiate. Tamalpais. Summer.

2. C. OREGANA Gray. Capitula not radiate.—Livermore. Autumn.

### 6. Aplopappus Cass.

Superflua. Æqualis. Head $\infty$-flowered. Ray-flowers ligulate. Bristles of the pappus $\infty$. Flowers yellow.

1. A. LINEARIFOLIUS DC. Rays more than 9.— ♄. Contra Costa mountains. Summer.

2. A. ERICOIDES Hook. Ray-flowers less than 10.— ♄. Sand hills. San Francisco. Summer.

### 7. Bigelovia DC.

Æqualis. Superflua by a single ray-flower. Head few-flowered. Receptacle narrow. Pappus 1-seriate; bristles $\infty$. Flowers yellow.

1. B. ARBORESCENS Gray. Scales of the involucre irregularly imbricate and not in distinct ranks.— ♄. Tamalpais. Summer.

## 8. Solidago L. GOLDEN ROD.

Superflua. Head $\infty$-flowered. Ray-flowers ligulate, few, distant from each other. Akenes terete, $\infty$-costate; pappus of $\infty$, capillary bristles.—♃. Autumn. Suffrutescent plants, with long rod-like branches. Flowers yellow.

1. S. OCCIDENTALIS Nutt. Stem branching. ♃. Moist places. Summer.

2. S. CALIFORNICA. Nutt. Stem single; the whole plant pubescent.—♃. Moist sand. Summer.

3. S. SEMPERVIRENS L. Stem simple; the plant completely glabrous.—♃. Salt marshes near San Francisco. Autumn.

All Solidago species seem to be possessed of diuretic powers. S. Virgaurea, a European species, formerly officinal, is still in use as a domestic remedy.

## 9. Corethrogyne DC.

Frustranea. Head $\infty$-flowered; ray-flowers $\infty$, 1-seriate; scales of the involucre imbricate; style appendages of disc-flowers with tuft-like bristles; pappus of simple, unequal rigid bristles.—♃. Ray-flowers blue, purple; habit of aster.

1. C. OBOVATA Benth. Contra Costa range. Marin County. Summer.

## 10. Aster (Nees.)

Superflua. Head ∞-flowered. Ray-flowers 1-seriate, ligulate. Scales of involucre imbricate. Style appendages triangular-lanceolate to subulate. Pappus of copious capillary, scabrous bristles. — ♃. Ray-flowers rose-color, white, blue.

1. A. RADULINUS Gray. Pappus rigid; some of the bristles thickened toward the top; rays white.—♃. Berkeley. Autumn.

2. A. CHAMISSONIS Gray. Pappus soft; stem glabrous, erect and branching, leafy; involucre imbricate, its scales with green tips; ray-flowers violet.—♃. San Francisco at the cemetery. Autumn.

3. A. DIVARICATUS Nutt. Pappus soft; stem glabrous, diffusely branched; branches slender; involucre imbricate, its scales with scarious margins.—☉. Salt marshes, San Francisco. Autumn.

## 11. Erigeron L.  FLEA-BANE.

Superflua. Necessaria. Head ∞-seriate, ligulate; scales of the involucre imbricate, linear. Pappus scanty of scabrous, capillary bristles.

1. E. STENOPHYLLUS Nutt. Stem leafy; leaves linear, scabrous, deep green; rays purple.—♃. Livermore. Autumn.

2. E. GLAUCUS Ker. Radical leaves differing from cauline, glaucous, entire, succulent; ray purple.—♃. Seashore. Summer. Autumn.

3. E. PHILADELPHICUS L. Leaves irregularly dentate; ray-flowers narrow, numerous, reddish.—♃. Colma. Berkeley. Autumn.

4. E. CANADENSIS L. Leaves numerous, small, rays inconspicuous, whitish.—☉. Waste grounds. Common. Summer. Autumn.

### 12. Baccharis L.

Diœcious. Head without rays. Scales of the involucre ∞-seriate, imbricate. ♂ corolla tubular with 5-cleft limb. Pappus capillary, 1-seriate. ♀ corolla filiform, truncate. Pappus copious, capillary.

1. B. PILULARIS DC. Leaves sessile cuneiform, sinuately dentate; flowers whitish.—♄. Summer. Autumn. Common.

2. B. DOUGLASII DC. Glutinous; leaves lanceolate, acute, distinctly 3-nerved; capitula in a compound, terminal corymb; receptacle conical.—♃. Moist sand. San Francisco. Summer.

3. B. VIMINEA DC. Shrub with the aspect of a willow; leaves lanceolate, acute; capitula in terminal corymbs and racemes, receptacle flat.—♄. River-bed, Niles. Livermore. Summer. Autumn.

## Tribe II. INULEÆ.

Anthers with tails, sagittate. Style branches without appendages.

### 13. Pluchea Cass.   MARSH-FLEABANE.

Superflua. Necessaria. Head ∞-flowered. ♀ flowers ∞-seriate, tubular. Scales of the involucre ∞-seriate, ovate, imbricate. Receptacle flat, naked. Pappus 1-seriate.

1. P. CAMPHORATA DC.  Flowers rose-color. —⊙. Salt marshes. Autumn.

### 14. Adenocaulon Hook.

Necessaria. Head few-flowered, all the flowers tubular, Anthers sagittate, not tailed. Scales of the involucre 5, 1-seriate, herbaceous, at last reflexed. Receptacle flat, naked. Akenes exserted, obovate, glandular near the summit; pappus 0.—♃. Flowers pale.

1. A. BICOLOR Hook.—Redwoods. Summer.

### 15. Micropus L.

Necessaria. Superflua. Head of few flowers, all tubular. ♂ 5, ♀ 5, embraced by the scales of the two-seriate involucre. Receptacle narrow, naked. Akenes falling off with the embracing scales of the involucre. Pappus 0. —⊙. Flowers inconspicuous.

1. M. CALIFORNICUS Fisch. & Meyer. Low woolly herb. Common. Spring.

## 16. Psilocarphus Nutt.

Necessaria. Superflua. Head ∞-flowered. ☿ flowers few, tubular, 5-dentate. ♀ flowers ∞-seriate, filiform, each wrapped in a palea. Scales of the involucre few, scarious. Receptacle convex, paleaceous externally, naked in the center. Akenes enveloped in the paleæ. Pappus 0.—⊙. Flowers inconspicuous.

1. P. OREGANUS Nutt. Covered with loose, white wool; akenes cylindrical.—Low ground near creeks. Spring.

2. P. TENELLUS Nutt. Covered with appressed grey wool. Akenes fusiform.—Low ground near creeks. Spring.

## 17. Evax Gærtn.

Necessaria. Superflua. Head ∞-flowered. ☿ flowers tubular, few— ♀ flowers ∞-seriate, filiform. Scales of the involucre few, 1-2-seriate, scarious. Receptacle elongate (like the axis of a spike), paleaceous externally, naked in the center with the ♂ flowers. Pappus 0.—⊙. Flowers inconspicuous.

1. E. CAULESCENS Gray.—Woolly annual. Gravelly alluvium. San Rafael. Spring.

## 18. Filago L.

Superflua. Head ∞-flowered. ☿ flowers tubular, 4-dentate. ♀ flowers ∞-seriate, filiform; external rows hidden between the scales of the involucre and the paleæ. Scales of the involucre imbricate, similar to the paleæ. Receptacle elongate (like the axis of a spike). Pappus of ∞, capillary bristles, depauperate towards the periphery. External row ♀; pappus pilose. ⊙. Flowers inconspicuous.

1. F. CALIFORNICA Nutt. Woolly annual. Common. Spring.

### 19. Anaphalis DC. PEARLY-EVERLASTING.

Necessaria. Superflua. Head ∞-flowered. ☿ tubular, 5-dentate; style scarcely divided, blunt; ♀ flowers ∞-seriate, filiform. Scales of the involucre imbricate, ∞-seriate, radiating; external rows ovate; internal rows longer and narrower. Receptacle flat, naked. Pappus 1-seriate, capillary, its bristles scabrous and distinct at base.—♃.

1. A. MARGARITACEA Benth. & Hook.—Common. Summer.

### 20. Gnaphalium L. EVERLASTING.

Superflua. Head ∞-flowered; ☿ flowers tubular, 5-dentate. Style 2-cleft. ♀ flowers filiform, ∞-seriate; involucre ovate. Scales imbricate, 8-seriate, scarious; as long as the head. Receptacle flat, naked. Pappus of 1-seriate, capillary bristles. Flowers pale.

1. G. MICROCEPHALUM Nutt. Bristles of pappus not united at their base into a ring; involucre mainly scarious, decidedly imbricate; leaves but slightly decurrent; white-woolly; involucre turbinate.—♃. Contra Costa.

2. G. SPRENGELII Hook. & Arn. Bristles of pappus not united at their base into a ring; involucre mainly scarious, decidedly imbricate; leaves but slightly decurrent, white-woolly; involucre hemispherical.—☉. Common. Summer.

3. G. DECURRENS Ives. Bristles of pappus not united at their base into a ring; involucre mainly scarious, decidedly imbricate; leaves decidedly decurrent, glandular, white-woolly only underneath; involucre companulate. Smell of the plant like liquorice.—☉. Sand-hills. San Francisco. Summer.

4. G. RAMOSISSIMUM Nutt. Bristles of pappus not united at their base into a ring; involucre mainly scarious, decidedly imbricate; leaves decidedly decurrent, linear, involucre turbinate. Smell of the plant like liquorice.—☉. Sand dunes. Saucelito. Summer.

5. G. PALUSTRE Nutt. Bristles of pappus not united at their base in a ring; scales of involucre woolly, only scarious at their tips, all nearly of the same length.—☉. Moist grounds. Common. Summer.

6. G. purpureum L. Bristles of pappus united at the base into a ring.—♃. Near salt marshes. Common. Summer.

### Tribe III. Ambrosiæ.

Anthers of ☿ or ♂ flowers distinct, not syngenetic. Style abortive, truncate. Corolla of ♀ flowers rudimentary or 0; pappus 0.

### 21. Iva L.

Necessaria. Superflua. Head ∞-flowered; ☿ or ♂ flowers ∞, tubular, 5-lobed; ♀ flowers few, 5-dentate. Scales of the involucre 3 to 4, ovate. Paleæ of the receptacle linear.—♃.

1. I. axillaris. Pursh.—Near the seacoast. Summer.

### 22. Ambrosia DC.   Ragweed.

Monœcious. ♂ head ∞-flowered; scales of the involucre united into a cup; receptacle flat, naked; ♀ head 1-flowered. Scales of the involucre united into a cup. Corolla 0. Akenes enclosed in the persistent involucre.

1. A. artemisiæfolia L. Leaves all bipinnatifid.—☉. Cultivated grounds. Summer.

2. A. psilostachya DC. Upper leaves pinnatifid, lower bipinnatifid.—♃. Cultivated grounds. Summer.

### 23. Franseria Cav.

Monœcious. ♂ head ∞-flowered. Scales of the involucre united into a cup. Paleæ of the flat receptacle filiform. ♀ head 1-, sometimes 2-4-flowered. Scales of the involucre united into a cup; if the cup be more than 1-flowered, forming as many cells as there are flowers. Involucre armed with ∞ rows of spines. Akenes enclosed in the persistent involucre.—♃.

1. F. BIPINNATIFIDA Less. Leaves 2-3-pinnately divided.—Sand dunes, San Francisco. Summer.

2. F. CHAMISSONIS Less. Leaves cuneate to ovate, obtusely serrate, the lower sometimes laciniate.—Sand dunes, San Francisco. Summer.

### 24. Xanthium Tourn. COCKLEBUR.

Monœcious. ♂ head ∞-flowered. Scales of the involucre 1-seriate, distinct. Receptacle cylindrical, paleaceous. ♀ head 2-flowered. Scales of the involucre united into a cup, armed with hooked spines. Corolla filiform. Akenes enclosed in the persistent involucre.—☉.

1. X. STRUMARIUM L. Leaves green on both sides, scabrous, irregularly serrate.—Roadsides. Common. Summer.

2. X. SPINOSUM L. Leaves whitish beneath, most of them deeply lobed.—Waste grounds. Common. Summer.

## Tribe IV. HELIANTHEÆ.

Anthers without tails. Receptacle paleaceous. Pappus not capillary.

a. *Paleæ persistent.*

### 25. Balsamorrhiza Hook. Nutt.

Superflua. Head ∞-flowered. Ray-flowers lanceolate. Receptacle flat, paleaceous. Pappus 0.—♃ Flowers yellow.

1. B. DELTOIDEA Nutt. Leaves with cordate base, entire to serrate.—Niles. Spring.
2. B. Hookeri. Nutt. Leaves pinnately or bipinnately parted.—Hillsides near Lake Chabot. Spring.

### 26. Wyethia Nutt.

Superflua. Head ∞-flowered; ray-flowers lanceolate. Scales of the involucre 2—3-seriate: external foliaceous; internal paleaceous. Receptacle flat, paleaceous. Pappus cup-shaped or aristate.—♃. Flowers yellow.

1. W. HELENIOIDES Nutt. Involucre foliaceous, spreading; pappus chaffy, not aristate; young plant tomentose; akenes pubescent towards the apex.—Dry hillsides. Mission Dolores. Spring.
2. W. GLABRA Gray. Involucre foliaceous, spreading; pappus chaffy, not aristate; plant glabrous, glutinous; akenes glabrous.—Tamalpais. Spring.

3. W. ANGUSTIFOLIA Nutt. Pappus aristate. Marin County. Spring.

### 27. Helianthella Torr. & Gray.

Frustranea. Head ∞-flowered. Receptacle flat; paleaceous. Akenes flattened; pappus represented by 2, marginal, aristæ, alternating with 2, caducous paleæ.—♃. Flowers yellow.

1. H. CALIFORNICA Gray.—Napa. Spring.

### 28. Helianthus L. SUNFLOWER.

Frustranea. Head ∞-flowered. Receptacle flat, paleaceous. Akenes slightly compressed; quadrangular. Pappus represented by 2, marginal caducous; paleæ with minute, intermediate ones. Ray-flowers yellow.

1. H. ANNUUS L. Receptacle flat; paleæ of receptacle 3-cleft; disc brown; lower leaves cordate, serrate.—⊙ Common. Summer.

2. H. SCABERRIMUS Benth. Receptacle flat; paleæ of receptacle entire, aristate, the awn as long as the disc-flowers; disc brown. ⊙. San Rafael. Summer.

3. H. EXILIS Gray. Receptacle flat; paleæ of receptacle entire, aristate, the awn longer than the disc-flowers; disc brown.—⊙. Lake county. Summer.

4. **H. Californicus** DC. Receptacle convex; paleæ of receptacle blunt, not aristate; disc-flowers yellow, only their anthers brown.—♃. Berkeley hills. San Rafael. Summer.

b. *Paleæ deciduous*.

### 29. Pugiopappus Gray.

Superflua. Head ∞-flowered. Ray-flowers broad, deeply crenate. Involucre 2-seriate. Pappus of the disc 0; of the ray 2-aristate. Flowers yellow. Leaves alternate.—⊙.

1. P. calliopsideus Gray.—Alma. Summer.

### 30. Leptosyne DC.

Superflua. Head ∞-flowered. Ray-flowers broad, deeply crenate. Involucre 2-seriate, Pappus 0. Flowers yellow.—⊙.

1. L. Stillmani Gray.—Alma. Summer.

### 31. Bidens L.    Beggar-Ticks.

Frustranea. Æqualis. Head ∞-flowered. Involucre 2-seriate. Pappus 2–4-aristate. Leaves opposite.

1. B. chrysanthemoides Michx. Flowers yellow.—⊙. Swamps near San Francisco (extinct) Marin county. Summer.

## 32. Madia Mol. Tar-weed.

Superflua. Necessaria. Æqualis. Ray-flowers ligulate, 3-dentate, scarcely longer than the involucre. Involucre 1-seriate, herbaceous; scales carinate, complicate, embracing the ray-akenes. Receptacle naked in the center, with 1-2 rows of paleæ between the disc and the ray. Akenes laterally compressed; pappus of the ray 0; of the disc usually 0. Viscid, resinous herbs. Flowers yellow.

1. M. Nuttallii Gray. Ray-flowers exserted and conspicuous; paleæ of the pappus fimbriate.—♃. Redwoods, Marin county. Spring,

2. M. radiata Kell. Ray-flowers exserted and conspicuous, obtusely 3-dentate; pappus 0.—☉. Antioch. Spring.

3. M. elegans Don. Ray-flowers exserted and conspicuous, acutely 3-lobed; pappus 0. ☉.—Common. Spring and summer.

4. M. sativa Mol. Ray-flowers short, inconspicuous; pappus 0; akenes of the disc 4-nerved, quadrangular.—☉. Common. Summer.

5. M. dissitiflora Gray. Ray-flowers short, inconspicuous; pappus 0; akenes of the disc without the four nerves and angles.—☉. Marin county. Summer.

6. M. FILIPES Gray. Ray-flowers short, inconspicuous; pappus 0; disc-flower solitary. ☉.—Antioch. Spring.

All the species covered by a resinous exudation of strong, generally disagreeable odor. Recommended by the old settlers in affections of the urinary organs.

### 33. Hemizonia DC.   TAR-WEED.

Necessaria. Superflua. Ray-flowers ligulate, dentate. Involucre 1-seriate, with concave scales embracing the ray-akenes. Receptacle flat, naked in the center, with a row of paleæ between disc and ray or sometimes throughout. Akenes of the ray convex externally, flat internally; pappus of the ray 0; pappus of the disc scaly, aristate, plumose or 0.—☉. Viscid, resinous herbs; flowers yellow, white.

1. H. LUZULÆFOLIA DC. Only the ray-akenes developed, these obovate, triangular; terminal area depressed; rays 3-lobed, frequently white.—☉. San Francisco. Summer.

2. H. MACRADENIA DC. Ray-flowers numerous with short ligulæ. Ray-akenes turgid, gibbose, the gibbosity pushing the terminal area to the inner angle so that the area appears lateral; receptacle conical; leaves not pungent; flowers always yellow.—☉. Tamalpais. Contra Costa. Autumn.

3. H. Parryi Greene. Ray-flowers numerous with short ligulæ; ray-akenes turgid, very gibbose, the gibbosity pushing the terminal area to the inner angle so that the area appears lateral, receptacle convex; bracts of involucre pungent; paleæ of receptacle not pungent; upper leaves short, acerose, lower pinnatifid; flowers yellow.—⊙. St. Helena. Summer.

4. H. pungens Torr. & Gray. Ray-flowers numerous with short ligulæ; ray-akenes turgid, very gibbose, the gibbosity pushing the terminal area to the inner angle so that the area appears lateral; receptacle convex; upper leaves and bracts of the involucre pungent; paleæ of the receptacle pungent also; lower leaves frequently bipinnatifid; flowers yellow. ⊙. Common. Summer.

5. H. corymbosa Torr. & Gray. Ray-flowers cuneate, more than 10. Ray-akenes turgid, very gibbose, the gibbosity pushing the terminal area to the inner angle so that the area appears lateral; akenes 4-5-nerved; receptacle flat, naked on the disc, but the disc flowers separated from the ray-flowers by a cup, formed by connate paleæ; ovary of the disc-flowers not developed; flowers yellow.—⊙. Common. Summer.

6. H. angustifolia DC. Ray-flowers cuneate, more than 10; ray-akenes turgid, very

gibbose, the gibbosity pushing the terminal area to the inner angle so that the area appears lateral; akenes 3-nerved, receptacle flat, naked on the disc, but the disc-flowers separated from the ray by paleæ, connate at their base. Flowers yellow.—☉. Berkeley. Tamalpais. Summer.

7. **H. truncata** Gray. Ray-flowers few, but with very large 3-lobed yellow ligula. Plant glabrous. ☉. Marin county. Summer.

8. **H. multiglandulosa** Gray. Ray-flowers few but with very large, 3-lobed ligula; plant hispid-glandular.—☉. Tamalpais. Summer.

Properties the same as *Madia*.

### 34. Holozonia Greene.

Necessaria. Superflua. Disc flowers ∞; ray-flowers 5 to 8, cuneiform, 3-cleft. Scales of the involucre corresponding in number to the ray-flowers, and completely embracing each akene. Receptacle flat, the paleæ united into a cup enclosing the disc-flowers. Pappus of the disc generally 0; of the ray hyaline, spreading —♃. Flowers white.

1. **H. filipes** Greene.—♃. Sonoma. Autumn.

### 35. Lagophylla Nutt.

Necessaria. Superflua. Ray-flowers 5, cuneiform, 3-lobed. Involucre of 5 scales, acuminate, with scarious margins, each completely enclosing the corresponding ray-akene. Receptacle flat, with 5 paleæ between the disc and ray. Pappus 0.—⊙. Flowers yellow, whitish.

1. L. RAMOSISSIMA Nutt. Pubescent.—⊙. Tamalpais. Summer.

2. L. CONGESTA Greene. Hispid.—⊙. Tamalpais. Summer.

### 36. Layia Hook. & Arn.

Superflua. Head ∞-flowered; ray-flowers cuneiform, 3-dentate. Involucre 1-seriate, herbaceous; scales acuminate, with scarious margins, completely enclosing the ray-akenes. Receptacle flat, with a row of paleæ between disc and ray, or paleaceous throughout. Ray-akenes linear, attenuate at the base, and with a flat area at the top; pappus of the disc various; of the ray 0.—⊙.

1. L. CARNOSA Torr. & Gray. Pappus bristles plumose; ray-flowers inconspicuous, small. Sands of sea shore. Marin County. Spring.

2. L. HETEROTRICHA Hook. & Arn. Pappus-bristles plumose, their hairs erect; ray-flowers large, canspicuous, 3-lobed, white.—Antioch. Livermore. Niles. Spring.

3. L. ELEGANS Torr. & Gray. Pappus-bristles plumose, their hairs woolly and interlaced; ray-flowers conspicuous, yellow.—Marin County. Spring.

4. L. HIERACIOIDES Hook. & Arn. Pappus bristles plumose, their hairs straight and erect; rays conspicuous, yellow, but little longer than the disc-flowers.—Marin County. Spring.

5. L. GAILLARDIOIDES Hook. & Arn. Pappus-bristles plumose; their hairs straight and erect; ray-flowers conspicuous, yellow; considerably longer than the disc-flowers. Contra Costa. Saucelito. Spring.

6. L. PLATYGLOSSA Gray. Pappus aristate; awns equal; ray-flowers yellow, frequently with white tips.—Saucelito. Spring.

7, L. CALLIGLOSSA Gray. Pappus aristate; awns unequal; ray-flowers yellow with white tips.—Berkeley. Spring.

8. L. CHRYSANTHEMOIDES Gray. Pappus 0; ray-flowers yellow with white tips.—Alameda. Milbrae. Spring.

### 37. Achyrachæna Schauer.

Superflua. Ray-flowers short and hidden. Involucre 1-seriate; scales with scarious margins, embracing the ray-akenes. Receptacle flat, naked in the centre, paleaceous towards the margin. Akenes clavate, those of the disc truncate, those of the ray with an epigynous disc; pappus of about 10 shining, silvery scales in two series, the outer considerably shorter than the alternate inner ones.—⊙. Flowers yellow or pale.

1. A. MOLLIS Schauer.—Common. Spring.

## Tribe V. HELENIEÆ.

Receptacle not paleaceous. Scales of the involucre herbaceous; pappus not capillary.

### 38. Coinogyne Less. (*Jaumea* Pers.)

Superflua. Involucre 2-seriate, imbricate; scales rounded. Receptacle conical, naked. Pappus 0.—♃. Flowers yellow, leaves opposite, fleshy.

1. C. CARNOSA Gray.—Salt marshes around San Francisco. Summer.

### 39. Burriellia DC.

Superflua. Ray-flowers few and short. Scales of involucre 5, ovate. Receptacle subulate, naked. Pappus of the ray 2-aristate (sometimes 0), as long as the corolla; of the disc 4-aristate.—☉. Flowers yellow. Leaves opposite.

1. B. MICROGLOSSA DC.—San Francisco, Spring.

### 40. Bæria Fisch. & Mey.

Superflua. Ray-flowers ovate, exserted. Scales of the involucre 10, ovate. Receptacle conical, naked, rough. Akenes angled or nerved; pappus paleaceous or aristate, sometimes 0. Flowers yellow. Leaves opposite.—☉.

1. B. MACRANTHA Gray. Pappus aristate, sometimes 0; akenes not quadrangular; recep-

tacle muricate-roughened; leaves 3-nerved, ciliate.—⊙. Marin County. Spring.

2. B. CHRYSOSTOMA Fisch & Mey. Pappus 0; akenes not quadrangular; receptacle muricate-roughened; leaves linear, not ciliate.—⊙. Common. Spring.

3. B. GRACILIS Gray. Akenes quadrangular; pappus uniform, aristate; receptacle muricate-roughened; plant hirsutely pubescent. Common. Spring.

4. B. CARNOSA Greene. Pappus uniform, aristate; akenes quadrangular; receptacle muricate-roughened; plant glabrous.—⊙. Salt marshes. Spring.

5. B. FREMONTII Gray. Pappus 4 awns, alternating with narrow, small paleæ; receptacle muricate-roughened. Besides the entire leaves, some palmate ones.—⊙. Contra Costa. Spring.

6. B. ULIGINOSA Gray. Pappus 2-3 awns, alternating with broad fimbriate paleæ (sometimes pappus altogether 0); receptacle muricate-roughened. Besides the entire leaves, some pinnately or bipinnately lobed ones.—⊙. Common. Spring.

### 41. Eriophyllum Lag (*Bahia* DC.)

Superflua. Ray-flowers ovate, exserted. Scales of the involucre lanceolate, united at the base; receptacle naked, alveolate. Pappus

4, 8, or 12 membranaceous scales. Flowers yellow.

1. E. STÆCHADIFOLIUM Lag. Frutescent; heads with short peduncles in loose cymes; receptacle alveolate.— ♄. San Francisco. Summer.

2. E. CONFERTIFLORUM Gray. Suffrutescent; Head with very short peduncles in compact cymes; receptacle not alveolate.— ♃. Common. Summer.

3. E. CÆSPITOSUM Dougl. Herbaceous; heads solitary, or few on large peduncles.—♃. Marin County. Summer.

### 42. Monolopia DC.

Superflua. Scales of the involucre united into a dentate cup. Receptacle conical, naked, papillate. Pappus 0. Flowers yellow.

1. M. MAJOR DC. Ray-flowers with 3–4 lobed ligulæ and an appendage on the opposite side; bracts of the involucre united.—☉. Oakland. Saucelito. Summer.

2. GRACILENS Gray. Ray-flowers with 3–4-lobed ligulæ and an appendage on the opposite side; bracts of the involucre distinct to the base.—☉. Santa. Cruz Mountains. Summer.

### 43. Lasthenia Cass.

Superflua. Scales of the involucre united into a dentate cup. Receptacle conical, naked, papillose. Akenes linear compressed. Pappus 5 to 10 firm scales (sometimes 0).—⊙. Flowers yellow.

1. L. GLABERRIMA DC. Pappus paleaceous. Near salt marshes. Spring.

2. L. GLABRATA Lindl. Pappus 0. Perfoliate.—Common. Spring.

3. L. CALIFORNICA Lindl. Pappus 0. Leaves clasping, not perfoliate.—Berkeley. San Mateo. Spring.

### 44. Rigiopappus Gray.

Superflua. Scales of the involucre linear, erect, rigid, half embracing the akene. Receptacle flat, naked. Pappus 3-5-aristate.—⊙. Flowers pale. leaves alternate.

### 45. Chænactis. DC.

Æqualis. Flowers of the ray tubular but frequently enlarged. Involucre hemispherical; its scales narrow. Receptacle flat. Pappus 4 to 12 obtuse, chaffy scales.—⊙.

1. C. LANOSA DC. Flowers yellow.—Contra Costa. Spring.

### 46. Blennosperma Less.

Necessaria. Ray flowers ovate, ligulate, without tube. Scales of the involucre 1-seriate, membranaceous. Receptacle flat, naked.

Akenes pyriform, papillate, gelatinous when wetted; pappus 0.—⊙. Flowers pale yellow. Leaves alternate.

1. B. CALIFORNICUM Torr. & Gray.—Common. Spring.

### 47. Helenium L.  SNEEZE-WEED.

Superflua. Flowers of the ray palmate. Involucre 2-seriate; external scales ∞, narrow, foliaceous, spreading, at length reflexed; internal scales few, paleaceous. Receptacle globular, naked. Pappus 5 to 12, membranaceous paleæ. Flowers of the ray yellow; disc purple; leaves alternate.

1. H. PUBERULUM DC.—⊙. Moist places. Summer.

### Tribe VI. ANTHEMIDEÆ.

Receptacle rarely paleaceous. Scales of the involucre scarious. Pappus not capillary.

### 48. Anthemis L.  MAY-WEED.

Superflua. Ray flowers ligulate. Scales of the involucre imbricate. Receptacle convex or conical, paleaceous. Akenes ribbed. Pappus 0. Ray-flowers white. Disc yellow.

I. A. COTULA L.—⊙. Common on waste grounds, although not indigenous. Native of Europe.

### 49. Achillea L. Yarrow.

Superflua. Head ∞-flowerad; ray-flowers 5 to 6, ligulate, ovate. Involucre ovate; scales imbricate, those of the receptacle hyaline. Akenes marginate; pappus 0. ♃.

1. A. MILLEFOLIUM L. Ray-flowers and disc pale.—Common. Summer.

### 50. Matricaria L. Chamomile.

Superflua. Æqualis. Scales of the involucre imbricate. Receptacle conical, naked. Akenes angulate. Pappus coroniform or 0. Disc yellow; flowers of the ray white.—☉

1. M. DISCOIDEA DC. Pappus 0; areola of akene merely surrounded by a rim; ray-flowers white, depauperate.—Summer. Common.

2. M. OCCIDENTALIS Greene. Pappus coroniform, unequal, the side toward the ray considerably developed, the inner side reduced to a rim.—Contra Costa. Summer.

### 51. Chrysanthemum L.

Superflua. Involucre hemispherical, spreading. Scales ∞-seriate, imbricate, appressed. Receptacle not conical, naked. Akenes short, somewhat terete, costate, truncate at the apex; pappus 0.

1. C. SEGETUM L. Ray-flowers yellow.—☉. Has been found near Oakland. Summer.

## 52. Soliva Ruiz & Pavon.

Necessaria. Superflua. Ray-flowers ∞, without corolla. Style scarcely bifid, persistent. Receptacle flat, naked. Akenes obcompressed, winged, crowned by the style, sessile, pappus 0.—⊙. Flowers inconspicuous.

1. S. SESSILIS Ruiz & Pavon.—Moist grounds near the coast. All the year round.

## 53. Cotula L. BRASS-BUTTONS.

Superflua. Corolla of ray-flowers wanting. Receptacle flat, naked, papillose. Akenes compressed, winged, those of the ray stipitate; pappus 0.—⊙. Flowers yellow.

1. C. CORONOPIFOLIA L. Leaves pinnatifid. Aquatic. All the year round.
2. C. AUSTRALIS Hook. Leaves bipinnatifid; lobes linear.—Waste places. All the year round.

## 54. Tanacetum L. TANSY.

Æqualis. Superflua. Ray-flowers not ligulate. Receptacle convex, naked. Akenes angulate, with large epigynous disc. Pappus coroniform, or 0.—♃. Flowers yellow.

1. T. CAMPHORATUM Less. Suffrutescent. Sand dunes near San Francisco. Summer.

## 55. Artemisia L. WORMWOOD. SAGE-BRUSH.

Æqualis. Superflua. Necessaria. Ray flowers tubular. Scales of involucre dry, with scarious margins. Receptacle naked or villous. Akenes obovate, with small epigynous disc; pappus 0. Heads small; flowers yellowish.

1. A. PYCNOCEPHALA DC. Superflua: flowers of the disc ☿, but sterile; their style not bifid; plant silky-villous; leaves pinnately parted.—♃. Seashore. Summer.

2. A. DRACUNCULOIDES Pursh. Superflua. Flowers of the disc ☿, but sterile; their style not bifid; plant glabrous; leaves linear.—♃. Lake Chabot. Summer.

3. A. CALIFORNICA Less. Superflua. Flowers of the disc ☿, and fertile; their style bifid; frutescent, paniculately branched, canescent; leaves pinnately parted; lobes filiform.—♄. Common. Summer.

Branches and leaves insecticide of considerable power.

4. A. LUDOVICIANA Nutt. Style of the flowers of the disc bifid; involucre tomentose; leaves lanceolate; the lower frequently parted into 3–5 lobes.—♃. Common. Summer.

Tribe VII. SENECIONIDÆ.

Receptacle naked. Pappus capillary.

56. **Nardosmia** Cass. (*Petasites*, Tourn.)

Monœcia, Necessaria. Heads dimorphous, ♂ or ☿ contains ♂ flowers ∞, ♀ flowers few. ♀ or ☿ head contains ☿ or ♂ flowers few, ♀ flowers ∞.

Involucre 1-seriate. bracteolate. Receptacle naked. Akenes glabrous, ribbed; pappus of copious capillary bristles —♃. Flowers pale, appearing before the leaves.

1. N. PALMATA Gray.—Taylorville. Santa Cruz Mountains. Spring.

### 57. Arnica L.

Superflua. Ray flowers elongate, with distinct, but sterile anthers. Involucre 2-seriate. Pappus capillary.—♃. Flowers yellow.

1. A. DISCOIDEA Benth.—Santa Cruz Mountains. Summer.

### 58. Senecio L.  GROUNDSEL.

Superflua. Æqualis. Involucre 1-seriate, bracteolate. Pappus capillary. Flowers yellow.

1. S. HYDROPHILUS Nutt. Glabrous; heads erect; rays few, sometimes 0; leaves almost entire; the radical, large, with long petioles; upper cauline leaves sessile, clasping.—♃ Berkeley. Summer.

2. S. ARONICOIDES DC. Heads erect; ray-flowers few, frequently 0; leaves repand-denticulate; cauline variable, uppermost reduced to bracts.—♃. Presidio. Spring.

3. S. EURYCEPHALUS Torr. & Gray. Rays elongated; leaves pinnately parted; lobes cuneate and acutely incised.—♃. San Mateo. Summer.

4. S. Douglasii DC. Ray-flowers elongated; leaves linear or pinnately parted into linear segments.—♃. Contra Costa hills. Summer.

5. S. vulgaris L. Rays 0.—☉. Waste grounds. All the year round. Weed introduced from Europe.

Tribe VIII. Eupatoriaceæ.
Style branches club-shaped. Æqualis.

**59. Bulbostylis** DC. (*Brickellia* Ell.)

Scales of involucre imbricate, 2–3-seriate. Receptacle flat, naked. Margin of the tubular corolla minutely 5-dentate. Style bulbous at base. Akenes 10-striate; pappus ∞-seriate, bristly, scabrous. Leaves opposite.

1. B. Californica Gray. Suffrutescent; flowers pale.—♃. Niles Station. Summer.

Tribe IX. Cynaroideæ.
Style branches concreted. Corolla deeply 5-cleft. Receptacle bristly.

**60. Cnicus** L.   Thistle.

Æqualis. Scales of the involucre ∞-seriate, imbricate, ending in a spine. Pappus deciduous, ∞-seriate; bristles plumose, connected at their base into a ring. Filaments syngenetic. ☉☉.

1. C. AMERICANUS Gray. Involucre ovoid; bracts appressed, imbricate, with loose, scarious tips; flowers ochroleucous.—⊙⊙. Marin County. Summer.

2. C. EDULIS Gray. Bracts of the involucre not appressed, not rigid, but loose, tapering from a narrow base gradually into a short flaccid spine; corolla lobes filiform, with a thickened tip; flowers purple or whitish.—⊙⊙. Common. Summer.

3. C. HALLII Gray. Bracts of the involucre not appressed, not rigid, but loose, tapering from a narrow base gradually into a flaccid, short spine; corolla lobes linear, flat; flowers rose-color or whitish.—⊙⊙. Marin County. Summer.

4. C. OCCIDENTALIS Gray. Bracts of the involucre with short coriaceous base and squarrose, subulate tips; heads solitary, terminal, very large; flowers crimson. Common. Summer.

5. C. FONTINALIS Greene. Bracts of the involucre herbaceous, broad, reflexed, with a short spinose tip; heads nodding; flowers whitish.—Crystal Springs; very local. Summer.

6. C. QUERCETORUM Gray. Bracts of the involucre appressed, coriaceous, only the outer mucronate, the inner unarmed; all plane without

glandular viscid spot on the dorsal side; flowers rose, purple or whitish.—Presidio. Oakland. Summer.

7. C. BREWERI Gray. Bracts of the involucre bearing on their dorsal side, near the tip, an oval, viscid, glandular spot; bracts much appressed, coriaceous, outer bracts tipped with a squarrose, slender prickle; flowers pale, purple or whitish.—Tamalpais. Summer.

### 61. Silybum Vaill. SPOTTED THISTLE.

Æqualis. Pappus $\infty$-seriate. Bristles rigid. Filaments monadelphous.—⊙. Flowers purple.

1. S. MARIANUM Gærtn.—Common. Summer.

Native of the Mediterranean region. Seeds of the plant used as an antispasmodic.

### 62. Centaurea L. STAR-THISTLE.

Frustranea. Æqualis. Involucre imbricate. Ray-flowers funnel-shaped. Pappus of $\infty$, scabrous, filiform bristles.

1. C. BENDICTA L. Akenes terete; flowers yellow.—⊙.

Ballast-weed. Introduced from the Mediterranean region. Summer.

2. C. MELITENSIS L. Akenes costate. Flowers yellow.—⊙.

Ballast-weed. Introduced from the Mediterranean region. Summer.

### Sub-Family 2. *Liguliflorœ*.
All flowers of the head ligulate.

#### Tribe I. CICHORIACEÆ.

**63. Stephanomeria** Nutt.

Head few-flowered. Involucre 1-seriate, calyculate. Receptacle flat, naked. Akenes 5-costate, with a callosity at their place of insertion. Pappus 1-seriate, bristly, plumose, white, coalescent.—☉.

1. S. VIRGATA Benth. Flowers matutinal. Common. Summer.

**64. Rafinesquia** Nutt.

Head ∞-flowered. Involucre 1-seriate, calyculate. Receptacle flat, naked. Akenes rostrate, with a callosity at their place of insertion; pappus white, plumose, capillary, coalescent.—☉. Flowers pale.

1. R. CALIFORNICA Nutt. Common. Summer.
The smell of the plant resembles that of opium.

**65. Hypochæris** L.

Involucre imbricate. Receptacle paleaceous. Paleæ deciduous. Akenes rostrate; pappus plumose. Flowers yellow.

1. GLABRA L.—☉. Ballast weed, introduced from Europe. Summer.

### 66. (*Microseris*) Scorzonella Nutt.

Involucre campanulate, imbricate in several ranks. Receptacle flat, naked, foveolate. Akenes more than 5-costate; pappus paleaceous; paleæ entire, lanceolate, tipped with a barbellate awn. Heads nodding when in bud. Plant caulescent. Flowers yellow.

1. S. PROCERA Gray. Head more than 100-flowered; awns simply barbellate, 3 times longer than the paleæ.—♃. Sonoma. Summer.

2. S. PALUDOSA Greene. Head less than 75 flowered; awns simply barbellate, twice as long as the paleæ.—♃. Marshy grounds. Marin County near Corte Madera. Summer.

3. S. SYLVATICA Gray. Awns nearly plumose, shorter than the paleæ.—♃. Contra Costa. Summer.

### 67. (*Microseris*) Calais DC.

Involucre conical, imbricate. Receptacle flat, alveolate. Akenes more than 5-costate; paleæ of the pappus 5, flat, bifid, with short awns. Head always erect on hollow, scape-like peduncles, that swell toward the head. Flowers yellow.

1. C. Kelloggii Greene. Awn of the pappus longer than the palea; palea emarginate (not deeply cleft).—⊙. San Bruno Mountains; Marin County. Spring.

2. C. Lindleyi DC. Awn of the pappus somewhat shorter than the palea; palea brownish, persistent.—⊙. Common. Spring.

3. C. linearifolia DC. Awn of the pappus not longer than the palea; palea white, persistent. ⊙.—Common. Spring.

### 68. (*Microseris*) **Microseris** Don.

Involucral bracts imbricate, the outer rank very short, calyculate. Receptacle flat, slightly alveolate. Paleæ of the pappus entire, tapering into a scabrous awn. Capitula nodding on slender scapes, which are not thickened towards the head. Flowers yellow.—⊙.

1. M. Douglassii Gray. Pappus paleaceous; paleæ 5, boat-shaped, tapering abruptly into an awn of the length of the akene.—Common. Spring.

2. M. attenuata Greene. Pappus paleaceous; paleæ 5, boat-shaped, tipped with an awn shorter than the akene.—Contra Costa. Spring.

3. M. acuminata Greene. Pappus paleaceous; paleæ 5, flat and straight, gradually tapering into an awn, shorter than the palea. Contra Costa. Spring.

4. M. Bigelovii Gray. Pappus paleaceous; paleæ 5, flat and straight, passing into an awn more than twice the length of the palea.—Common. Spring.

5. M. elegans Greene. Pappus paleaceous; paleæ deltoid, very short, with a slender awn twice their length.—Niles. Spring.

6. M. aphantocarpha Gray. Pappus paleaceous; paleaceous part nearly obsolete, aristate part long but setaceous, slender, and very fragile.—San Francisco. Spring.

### 69. Malacothrix DC.

Receptacle, flat, naked. Akenes truncate; pappus 1-seriate, bristles scabrous, silvery, deciduous in a ring.—☉.

1. M. Californica DC. Scapose; involucral bracts in more than 2 ranks; flowers yellow. Contra Costa. Spring.

2. M. Clevelandi Gray. Branching; involucral bracts in 2 ranks of equal length; besides the regular caducous pappus an external accessory pappus of 1 persistent bristle and a dentate crown; flowers yellow.—Contra Costa. Spring.

3. M. obtusa Benth. Branching; involucral bracts in two ranks of unequal length; persistent bristles 0; flowers white.—Marin County. Spring.

### 70. Hieracium Tourn. HAWKWEED.

Involucre ∞-seriate, imbricate. Akenes terete, with a thin, crenulated ring on the apex, pappus 1-seriate, capillary.—♃.

1. H. ALBIFLORUM Hook. Involucre ∞-flowered; akenes linear; pappus sordid; flowers white.—Saucelito. Summer.

### 71. Troximon Nutt.

Involucre imbricate. Akenes linear. Apex contracted. Pappus ∞-seriate. Bristles silky, not deciduous in a ring.

1. T. APARGIOIDES Less. Beak of the akene about the length of body; flowers yellow.—♃. San Francisco. Summer.

2. T. HUMILE Gray. Beak of the akene several times the length of body; pappus as long as the beak; flowers yellow.—♃. Common. Summer.

3. T. LACINIATUM Gray. Beak of the akene several times the length of the body; pappus much shorter than the beak; leaves laciniate—when more deeply parted and pinnatifid, the lobes linear; flowers yellow.—♃. Marin County. Summer.

4. T. GRANDIFLORUM Gray. Beak of the akene several times the length of body; pappus much shorter than the beak; leaves runcinate

—when more deeply parted and pinnatifid, the lobes not linear; flowers yellow.—♃. Common. Summer.

5. T. HETEROPHYLLUM Greene. Beak of the akene but little longer than body; flowers yellow.—☉. Common. Spring.

### 72. Taraxacum Haller. DANDELION.

Involucre imbricate. Akenes contracted into a long beak; pappus ∞-seriate; bristles white, capillary nearly persistent.—♃. Flowers yellow.

1. T. OFFICINALE Weber. Ballast weed introduced from Europe; rare. Summer. Officinal under the old Linnean name. *Leontodon Taraxacum.*

### 73. Sonchus L. SOW-THISTLE.

Involucre ∞-seriate, imbricate. Akenes compressed, truncate. Pappus ∞-seriate, capillary. Hairs in fascicles, soft. Flowers yellow.

1. S. OLERACEUS L. Auricles of the cauline leaves acute.—☉. Common. All the year round. Ballast weed from Europe.

2. S. ASPER Vill. Auricles of the clasping cauline leaves rounded.—☉. Ballast weed from Europe.

## Family 2. CAMPANULACEÆ.

Calyx superior, 5-cleft. Stamens 5, alternating with the divisions of the gamapetalous corolla, and inserted between the corolla and the ovary. Anthers introrse, filaments flattened. Ovary 2–5 celled; cells $\infty$-ovulate. Placentæ central. Style 1. Stigmas 2–5. Fruit capsular. Leaves alternate.—Lactescent herbs.

### 1. Githopsis Nutt.

Calyx tube clavate, 10-costate; with 5 narrow, persistent lobes. Corolla companulate, 5-lobed. Filaments 5, membranaceous. Ovary 3-celled. Stigmas 3. Capsule clavate, 10-costate, enclosed by the calyx, opening by a round hole at the apex. Seeds $\infty$.—$\odot$. Dwarfish.

1. G. SPECULARIOIDES Nutt. Flowers blue. Marin County. Contra Costa. San Francisco. Spring.

### 2. Specularia Heister. VENUS'S LOOKING-GLASS.

Calyx-tube elongated. Lobes narrow. Corolla rotate, 5-lobed. Filaments membranaceous. Capsule elongated, opening by parietal valves near the apex. Seeds $\infty$.—$\odot$. Flowers blue.

1. S. BIFLORA Gray.—Marin County. Contra Costa. Spring.

### 3. Campanula Tourn.    BELL-FLOWER.

Calyx tube turbinate. Corolla campanulate. Filaments membranaceous at their base. Capsule turbinate, opening by parietal valves. Flowers blue.

1. C. EXIGUA Rattan. Stem-leaves linear.—☉. Tamalpais. Spring.

2. C. PRENANTHOIDES Durand. Stem-leaves ovate, serrate.—♃. Marin County redwoods. Spring.

### 4. Heterocodon Nutt.

Early flowers cleistogamic; later ones expanding. Calyx turbinate. Corolla campanulate. Capsule membranaceous, turbinate, opening irregularly.—☉. Flowers blue.

1. H. RARIFLORUS Nutt.—Marin County. Spring.

### Family 3. CUCURBITACEÆ.

Monœcious or dioecious. ♂: Stamens 5, 3-adelphous. Cells of anthers contorted, flexuous. ♀: ovary 2-5-celled. Placentæ parietal. Style 1. Stigmas 3.

### 1. Megarrhiza Torr. (*Echinocystis.* Torr. & Gray.)
BIG-ROOT.

Monœcious. ♂ flowers racemose. ♀ flowers single from the axil of the ♂ raceme. Limb of calyx minutely 5-dentate. Corolla rotate, deeply 5-lobed. Fruit a berry. Seeds large, globose.—♃. Root large; stem climbing by tendrils. Flowers greenish white.

1. M. FABACEA Naud. (*Californica* Torr.) Ovary globose, densely echinate; fruit globose, covered with pungent spines; seeds globose.—San Francisco. Spring.

2. M. MARAH Kellogg. Ovary ovate, covered with soft spines; fruit ovate, muricate; seeds lenticular.—San Francisco. Spring.

The enormous rhizomes of both species abound in an acrid, drastic juice, which renders the fecula, contained in a fair proportion, poisonous.

Order 2. RUBIALES. Filaments terete.

### Family 1. DIPSACEÆ.

Leaves opposite. Stipules 0. Flowers in involucrate heads; single flowers protected by a double (involucellate) calyx, tetrandrous. Ovary 1-celled, 1-ovulate. Fruit utricular.

#### 1. Dipsacus Tourn.    TEASEL.

Head oblong. Leaflets of the involucre radiate. External calyx (involucellum) quadrangular; internal cyathiform.—⊙⊙. Spring.

1. D. FULLONUM L. Flowers pale.—Cultivated grounds. Summer. Escaped from cultivation. Originally Mediterranean.

## Family 2. VALERIANACEÆ.

Leaves opposite. Stipules 0. Inflorescence cymose. Calyx pappus-like. Ovary 3-celled; 2 cells sterile, and the fertile one 1-ovulate. Fruit an akene.

### 1. Plectritis. DC.

Limb of c a l y x straight, entire, obsolete. Tube of the corolla gibbous; limb 5-cleft, 2-labiate; triandrous. Akene winged by the persistent, sterile cell.—☉.

1. P. MACROCERA Gray. Corolla almost regular; a k e n e semilunar; dorsal angle of akene obsolete; flowers white or rose-color. Common. Spring.

2. P. CONGESTA Lindl. Corolla distinctly 2-labiate; dorsal angle of akene distinctly carinate; fl o w e r s rose-color.—Marin County. Spring.

## Family 3. CAPRIFOLIACEÆ.

Leaves opposite. Stipules 0. Fruit baccate.

### 1. Sambucus Tourn. ELDER.

Limb of Calyx 5-dentate. Corolla rotate, regular. Ovary 3-5-celled; cells 1-ovulate: styles 9; stigmas 3-5. Fruit a 3-5-seeded berry.

1. S. GLAUCA Nutt. Flowers cream-colored; berries blue, pruinose.— ♄. Berries eatable; root a powerful sudorific.

## 2. Symphoricarpus Cass.   SNOWBERRY.

Calyx-limb 5-4-dentate, persistent. Corolla funnel-shaped, 5-4-lobed. Ovary 4-celled; 2 cells ∞-ovulate but abortive, alternating with 1-ovulate and fertile. Style filiform. Berry 4-celled, 2-seeded.— ♄.

1. S. RACEMOSUS Michx. Glabrous; leaves oval; corolla narrowed at base; flowers pinkish; berries white.—Common. Summer.

2. S. MOLLIS Nutt. Pubescent; leaves orbicular; corolla with broad base; flowers pinkish; berries white.—Common. Summer.

## 3. Lonicera L.   HONEYSUCKLE. TWIN-BERRY.

Calyx-limb small, 4-dentate. Corolla tubular or campanulate. Limb irregular 5-cleft. Style filiform. Berry 3-celled; cells few-seeded.

1. L. INVOLUCRATA Banks. Flowers in pairs; bracts foliaceous, bractlets conspicuous and accrescent in fruit; flowers orange-colored; berries shining black, enveloped in dark red, involucrate bracts. Stem erect.— ♄. Common. Summer. "Twin-berry"

2. L. HISPIDULA Dougl. Flowers in whorls round an axis, forming a loose spike; uppermost leaves connate; flowers in different shades of red and yellow; berries orange-red. Stem not erect.— ♄. Common. Summer.

### Family 4. RUBIACEÆ.

Leaves opposite or verticillate. If opposite, stipulate.

#### 1. Cephalanthus L.   BUTTON-BUSH.

Flowers capitate. Limb of the calyx 4-dentate. Corolla tubular, slender, with 4-cleft limb. Stamens 4, inserted in the throat of the corolla. Style 1, exserted. Stigma capitate. Ovary 2-4 celled. Cells 1-ovulate.— ♄.

1. C. OCCIDENTALIS L. Flowers white; rare. Niles Station. Summer.

#### 2. Galium L.   BED-STRAW.  CLEAVERS.

Limb of the calyx entire, obsolete. Corolla 4-3 cleft, rotate. Ovary 2-celled. Cells 1-ovulate. Styles 2, connate at base, capitate at apex. Fruit sometimes fleshy. Stems quadrangular. Leaves verticillate.

1. G. APARINE L.— ☿  Fruit a dicoccous schizocarp, granulate and setaceous; leaves 6-8 in a whorl; stem retrorsely bristly; flowers white.—☉. Common. Summer.

2. G. BOREALE L.— ☿. Fruit a dicoccous schizocarp; leaves 4 in a whorl, 3-nerved, not cuspidate; flowers white.— ♃. Tamalpais. Summer.

3. G. TRIFIDUM L.— ☿. Fruit a dicoccous schizocarp; leaves 4-6 in a whorl, 1-nerved, not cuspidate; flowers small, white.— ♃. Marin County. Summer.

4. G. ASPERRIMUM Gray.— ☿ Fruit dicoccous, a scabrous schizocarp; leaves 4-6 in a whorl, cuspidate; cymes dichotomous; flowers white.—♃. Common. Summer.

5. G. TRIFLORUM Michx.— ☿ Fruit dicoccous, an uncinate hispid schizocarp; leaves 4-6 in a whorl, cuspidate; flowers in 3-florous or biternate cymes, whitish.—♃.

In shady, moist places; Marin County. San Mateo. Summer. This plant developes, in withering, the fragrance peculiar to the European *Asperula odorata*, the *Waldmeister* of the Germans. It is used frequently for the so-called "May-drink."

6. G. CALIFORNICUM Hook & Arn. Dioecious. Fruit baccate; leaves thinnish, ovate, acuminate, 4 in a whorl; flowers yellowish.—♃. San Francisco. San Mateo. Contra Costa. Summer.

7. G. NUTTALLII Gray. Dioecious. Fruit baccate; leaves coriaceous, small, oval, obtuse, mucronulate, 4 in a whorl; flowers yellowish; ♄.—Common. Summer.

8. G. ANDREWSII Gray. Dioecious. ♂ flowers in a few flowered cyme; ♀ flowers solitary; fruit baccate; leaves crowded, acerose subulate; 4 in a whorl.—♃. Common. Summer.

## Sub-Section 2. HYPOGYNÆ.

ORDER 1. **GENTIANALES.** Tendency to develop an apocarpus ovary either by separating the two cells entirely or by making the capsule septicidous.

### Family 1. ASCLEPIADACEÆ.

Monadelphous. Gynandrous. Anthers extrose; their pollen changed into pollinaria. Ovaries 2. Styles 2, united into a pentagonal stigma. Fruit a pair of follicles, or by abortion a single one. Lactescent plants.

#### 1. **Asclepias** L.   MILK-WEED.

Corolla rotate, 5-parted. Stamineal tube short, expanding into a corona of 5 cucullate processes, from the cavity of each of which rises a corniform process. Cells of anthers separated, each connected to the cell of the neighboring one by an apical appendage, the pollinia becoming suspended over a gland, five of which protrude from the margin of the stigma. Seeds ∞, with a silky arillus.—♃.

1. A. SPECIOSA Torr. Follicles echinate; flowers purplish.—Marin County. Contra Costa. Laundry Farm. Summer.

2. A. VESTITA Hook. & Arn. Follicles glabrous; erect on a deflexed pedicel; umbels ∞-flowered, terminal umbel pedunculate; lateral

ones sessile; flowers whitish or purplish.—Niles Station. Summer.

3. A. MEXICANA Cav. Follicles glabrous; erect on erect pedicels; umbels ∞-flowered; peduncles longer than the pedicels; flowers whitish.—Alameda on marshy places. Summer.

### Family 2. APOCYNACEÆ.

Stamens not monadelphous. Anthers introse. Corolla contorted in æstivation and its lobes oblique. Lactescent.

1. **Apocynum** Tourn. DOG-BANE. INDIAN HEMP.

Corolla companulate, 5-cleft, with 5 scales opposite the lobes and near their base. Stamens inserted on the base of the corolla. Filaments very short. Anthers sagitate, conniving into a ring. Ovaries 2, with 5, hypogynous scales. Fruit 2-follicular. Seeds with a hairy arillus.—♃.

1. A. ANDROSÆMIFOLIUM L. Cymes loose; corolla campanulate; tube longer than the calyx-lobes; corolla lobes revolute; flowers flesh-color.—Livermore. Summer. Rare. (Dogbane.)

2. A. CANNABINUM L. Cymes dense, compact; corolla tube not longer than the calyx-lobes; corolla lobes almost erect; flowers pale. Alvarado, on marshy places; very local. Sum-

mer. (Indian hemp.) The strong, silky fiber of the stem has been recommended as a textile substance, comparable to hemp.

### Family 3. GENTIANACEÆ.

Anthers introrse. Lobes of the corolla not oblique. Ovary ∞-ovulate, septicidal. Not lactescent.

#### 1. Microcala Link. (*Cicendia* Adans.)

Calyx costate, 4-dentate. Corolla salver-shaped. Withering anthers not spirally contorted. Style filiform. Stigma peltate or 2-lobed.—⊙. Dwarfish herbs. Flowers yellow.

1. M. QUADRANGULARIS Gries. Saucelito. Spring.

#### 2. Erythræa Pers.  CANCHALAGUA.

Calyx tubular, angular, 5-cleft. Corolla funnel-shaped, withering, but not deciduous. Withering anthers spirally contorted. Style filiform, deciduous. Capsule 1-celled, half divided by the prominent parietal placentæ. ⊙. Flowers pink.

1. E. FLORIBUNDA Benth. Anthers oblong; lobes of corolla lanceolate; seeds globular. San Bruno range. Spring.

2. E. MUHLENBERGII Gries. Anthers oblong; lobes of corolla very obtuse; seeds oval. San Miguel. Spring.

3. E. Douglasii Gray. Anthers oblong; lobes of corolla obtuse; seeds globular.—San Francisco. Spring.

4. E. trichantha Gries. Anthers linear; lobes of corolla narrow, lanceolate.—Marin County. Spring.

5. E. venusta Gray. Anthers linear; lobes of corolla oval.—San Leandro. Spring.

### 3. Gentiana L.  Gentian.

Corolla campanulate 4-5 lobed, withering but not deciduous. Style short or 0. Stigma 2-parted, persistent. Capsule septicidal.—♃.

1. G. Oregana Engelm. Flowers blue.—Tamalpais. Very local. Autumn.

### 4. Menyanthes Tourn.  Buck-bean.

Corolla funnel-shaped, induplicate in æstivation, deciduous. Limb 5-lobed, bearded on the upper surface.—♃. Aquatic. Flowers white.

1. M. trifoliata L. San Francisco. Summer. (Extinct since 1859.)

Order 2—DIANDRÆ. Corolla regular. Gamopetalous, 4-5 cleft or 0. Stamens 2. Ovary 2-celled; ovules in each cell 1-3. Leaves opposite.

### Family 1. OLEACEÆ.

Æstivation valvate.

### 1. Fraxinus Tourn.  ASH.

Flowers diclinous. Calyx 2–4 parted or 0. Corolla 2–4 parted; divisions 2–4 or 0. Ovary 2-celled; cells 2-ovulate. Stigma sessile, 2-cleft. Fruit a samara.— ♄.

  1. F. DIPETALA Hook. & Arn. Petals 2. ♄.—Livermore. Niles. Spring.

  2. F. OREGANA Nutt. Diœcious, apetalous. ♄.—Near creeks. Marin County. Menlo Park. San Mateo. Niles. Spring.

Infusion of leaves has formerly been used at the Russian settlements against rheumatism.

ORDER 3—LAMIALES. Corolla 2-labiate. Number of stamens less than lobes of the corolla. Ovaries 2–4 or ovary 2–4 celled. Cells 1–2 or 4-ovulate. Fruit an akene, drupe or capsule. Leaves opposite, Stipules 0.

### Family 1. LABIATÆ.

Ovary 4-cleft, 4-ovulate, inserted on a hypogynous disc. Style 1. Fruit 4 akenes.

### 1. Trichostema Gronov.  BLUE CURLS.

Tube of the corolla slender, limb 5-cleft. Lobes oblong, declined. Stamens didynamous, long-exserted. Filaments spiral in the bud.

  1. T. LANCEOLATUM Benth. Flowers blue. ☉.—Niles Station. Livermore. Dry river beds. Summer.

## 2. Mentha L.  Mint.

Calyx 5-dentate. Tube of corolla included. limb exserted, 4-lobed; upper lobe the largest. Stamens 4, scarcely didynamous, erect, distant. Anthers 2-celled. Cells parallel.—♃.

1. M. VIRIDIS L. Inflorescence terminal; dense glomerules, crowded around narrow spikes; spikes leafless; leaves almost sessile. Irrigated grounds. Summer. Autumn. Escaped from cultivation. Native of Europe. (Spearmint.)

2. M. PIPERITA L. Inflorescence terminal; glomerules on uninterrupted narrow spikes; spikes leafless; leaves distinctly petioled; flowers pale.—Irrigated grounds. Summer. Autumn. Escaped from cultivation. Native of Europe. (Peppermint.)

3. M. CANADENSIS L. Inflorescence axillary; uppermost axils of leaves without flowers.—Wet places. Native species. Summer.

## 3. Lycopus Tourn.  Water Horehound.

Diandrous. Posterior pair of stamens 0 or sterile. Otherwise as Mentha.—♃.

1. L. LUCIDUS Turcz. Flowers pale—Marin County. Aquatic. Summer.

## 4. Pycnanthemum Benth.  Mountain Mint.

Corolla tube as long as the calyx, limb 2-labiate, upper lip nearly entire. Lower lip

3-lobed. Lobes obtuse. Stamens 4, straight, divergent. Cells of anthers parallel.—♃.

1. P. CALIFORNICUM Torr. Flowers pale. Contra Costa. Summer.

### 5. Monardella Benth.

Limb of corolla 2-labiate. Upper lip 2-cleft. Lower lip 3-cleft. All the lobes narrow. Stamens 4, straight, divergent. Cells of anthers at last divergent or divaricate.

1. M. VILLOSA Benth. Flowers capitate; leaves ovate, pinnately veined. Bracts ovate, foliaceous; flowers purplish.—♃. Marin County. Summer.

2. M. DOUGLASII Benth. Flowers capitate; leaves lanceolate; bracts ovate, pinnately veined, membranaceous, tapering gradually into a cusp; flowers purplish.—⊙. Marin County. Summer.

3. M. BREWERI Gray. Flowers capitate. leaves ovate, pinnately veined, abruptly cuspidate; flowers purplish.—⊙. Contra Costa. Summer.

4. M. LANCEOLATA Gray. Flowers capitate; leaves lanceolate; bracts nerved, with veinlets crossing from nerve to nerve; flowers rose-color.—⊙. Contra Costa. Summer.

5. M. UNDULATA Benth. Flowers capitate; leaves spathulate, obtuse, undulate-margined; bracts ovate, nerved, without veinlets crossing

from nerve to nerve; flowers purplish.—☉. Common. Summer.

### 6. Micromeria Benth. YERBA BUENA.

Calyx tubular, 13-nerved, 5-dentate. Corolla 2-labiate. Stamens didynamous, ascending.—♃.

1. M. DOUGLASII Benth. Flowers white. Common. Summer.

Formerly much in use as a carminative.

### 7. Pogogyne Benth.

Calyx campanulate, 15-nerved, 5-cleft. 2 lower teeth much longer than the three upper ones. Corolla 2-labiate. Stamens didynamous, ascending, convergent in pairs. Style villous.—☉.

1. P. DOUGLASII Benth. All stamens fertile; inflorescence taking the shape of a continuous spike; lower calyx lobes twice the length of the calyx-tube; flowers violet.—Common. Summer.

2. P. PARVIFLORA Benth. All stamens fertile; inflorescence taking the shape of a continuous spike; lower calyx lobes about the length of the calyx-tube; flowers violet.—Berkeley hills. Summer.

3. P. SERPYLLOIDES Gray. Posterior pair of stamens sterile; inflorescence taking the shape of an interrupted spike; flowers violet. Berkeley. Summer.

### 8. Acanthomintha Gray.

Calyx tubular 13-nerved, naked in the throat. Upper lip 3-dentate. Lower lip 2-parted, shorter. Teeth spinulose. Corolla tube exceeding the calyx, naked. Upper lip curved, entire or 2-lobed. Lower lip 3-lobed. Upper pair of stamens sterile and shorter. Lower pair ascending. Anthers 2-celled. Cells divaricate. Bracts thorny.—⊙.

1. A. LANCEOLATA Curran. Flowers rose-color.—Niles. Summer.

### 9. Sphacele Benth.

Calyx campanulate, enlarged in fruit, 10-nerved, reticulate-veined, 5-dentate. Tube of the corolla wide, with a hairy ring near its base, the 5 lobes of the limb obtuse. Lower lobes the longer ones. Stamens ascending. Cells of the anthers divergent.— ♄. Flowers large, pale.

1. S. CALYCINA Benth.—San Mateo. Summer.

### 10. Salvia L. SAGE.

Calyx 2-labiate. Corolla ringent. Stamens 2, parallel. Anthers separated by a connective, shaped like a branch of the filament; only the posterior end of it bearing an anther cell.

1. S. CARDUACEA Benth. White-woolly with cobwebby hair; radical leaves sinuate to

pinnatifid, spinulose; upper lip of calyx 3-dentate; flower lavender-color.—☉. Livermore. Spring.

2. S. COLUMBARIA Benth. Puberulent; radical leaves pinnately parted; partitions crenate, rugose; upper lip of calyx 2-dentate, teeth connivent; flowers blue.—☉. Niles station. Spring.

### 11. Audibertia Benth.

Calyx 2-labiate. Corolla 2-labiate; upper lip 2-cleft, spreading. Stamens ascending. Anthers dimidiate, (upper half of the apparent filament showing by an articulation or spur its being originally a connective between the 2 cells of an anther.)

1. A. GRANDIFLORA Benth. Leaves hastate, sticky; flowers crimson.—♃. ♄. Bernal Hights. Summer.

2. A. STACHYOIDES Benth. Leaves lanceolate; flowers violet.—♄. Lake Chabot. Summer.

### 12. Lophanthus Benth.  GIANT HYSSOP.

Calyx subregular, 15-nerved, 5-dentate. Upper pair of stamens longer and declined; lower pair shorter and ascending. Anther 2-celled; cells parallel.—♃. Flowers purplish.

1. L. URTICIFOLIUS Benth.—Marin County. Summer.

### 13. Scutellaria L.  SKULLCAP.

Calyx 2-labiate, closed in fruit, at last splitting and the upper lip deciduous. Tube of corolla exserted; upper lip crest-shaped, with 2 lateral lobes; lower lip reduced to a single lobe. Stamens didynamous, ascending, parallel. Anthers of the lower pair dimidiate.

1. S. TUBEROSA Benth. Root stock filiform, bearing tubers; none of the floral leaves shorter than the flowers; flowers blue.—♃. San Francisco. Contra Costa. Spring.

2. S. CALIFORNICA Gray. Root stock filiform, but without tubers; uppermost floral leaves shorter than the flowers; flowers pale.—♃. Marin County. Spring.

### 14. Prunella L.  SELF-HEAL.

Calyx 2-labiate, closed in fruit and flattened. Corolla with a scaly or hairy ring near its base; upper lip concave; lower lip 3-lobed. Stamens didynamous, ascending. Anthers approximate in pairs.—♃.

1. P. VULGARIS L. Flowers blue.—San Francisco. Summer.

### 15. Marrubium L.  HOREHOUND.

Calyx tubular, 5-10 dentate. In fruit the teeth spreading. Tube of the corolla included. Limb 2-labiate. Stamens didynamous, included. Cells of anthers divaricate.—♃.

1. M. VULGARE L. Flowers white.—San Francisco. Contra Costa. Summer.

Native of Europe. Perhaps in older times used as a medicinal herb.

### 16. Stachys.  HEDGE-NETTLE.

Limb of the corolla 2-labiate, upper lip erect, concave, lower lip spreading, 3-lobed, middle lobe longest. Stamens ascending, approximate in pairs, parallel, contorted in withering.

1. S. AJUGOIDES Benth. Corolla pale, tube about the length of the calyx; leaves oblong with rounded base; lower leaves petioled; upper sessile; none of the floral leaves shorter than the clusters, which are arranged in an interrupted spike.—♃. Common. Summer.

2. S. ALBENS Gray. Corolla pale, tube about the length of the calyx; leaves oblong with cordate base; lower leaves short-petioled; upper nearly sessile; most of the floral leaves shorter than the inflorescences, which are arranged in an interrupted spike.—♃. Santa Clara County. Summer.

3. S. PYCNANTHA Benth. Corolla pale; tube about the length of the calyx; floral leaves all reduced to small bracts.—♃. Common. Summer.

4. S. BULLATA Benth. Corolla purple; tube somewhat longer than the calyx.—♃. Common. Summer.

5. S. CHAMISSONIS Benth. Corolla red; tube twice as long as the calyx.—♃. Presidio. Saucelito. Wild Wood Glen. Point Bonita. Summer.

### Family 2. VERBENACEÆ.

Didynamous. Ovary undivided. Fruit drupaceous or baccate, or splitting at length into its component parts (nutlets.)

#### 1. Verbena L.

Calyx 5-cleft. Corolla salver-shaped. Fruit at length splitting into four 1-seeded nutlets.

1. V. OFFICINALIS L. Flowers pale.—⊙. Not common. Summer.

#### 2. Lippia L.

Fruit 2-celled, splitting into two 1-seeded nutlets.

1. L. NODIFLORA Michx. Flowers rose-color. ⊙. Niles. Summer.

### Family 3. PLANTAGINEÆ.

Calyx 4-sepalous, persistent. Corolla gamopetalous, 4-cleft, scarious. Stamens 4, alternate with the lobes of the corolla. Anthers versatile. Style 1.

## 1. Plantago L. Plantain.

Flowers bracteate, spicate. Ovary 2-celled (spuriously 4-lobed.) Fruit a capsule, transversally dehiscent.

1. P. MAJOR L. Corolla glabrous; leaves 5–9 nerved; cells of ovary containing several ovules.—♃. Everywhere. Summer.

Popular remedy with the old inhabitants. Leaves used as external application; supposed to cool when put to a sore with their underside, and to draw when put there by their upper side.

2. P. MARITIMA L. Corolla tube externally pubescent; leaves linear, fleshy.—♃. On rocks near the sea. Summer.

3. P. LANCEOLATA L. Corolla glabrous; leaves 3–5 nerved; cells of ovary each containing 1 ovule.—⊙⊙. San Mateo. Summer.

4. P. PATAGONICA Jacq. Entire inflorescence silky pubescent; leaves 1–3-nerved; cells of ovary each containing 1 ovule.—⊙. Everywhere. Spring.

5. P. HIRTELLA HBK. Dioecious; scape and inflorescence hirsute; leaves 5–7 nerved; cells of ovary each containing 2 ovules.—♃. Ocean Lake. Berkeley. Summer.

An infusion of the leaves is much recommended by old inhabitants as a gargle in diphtheria.

6. P. Bigelovii Gray. Polygamous, diandrous; spikes short, thick, dense; leaves filiform, fleshy.—☉. Salt marshes. Summer.

7. P. heterophylla Nutt. Diœcious; spike slender; leaves fleshy.—☉. Sandy ground. Spring.

Order 4. **PERSONALES.** Corolla gamopetalous, irregular, its lobes not corresponding in number to the number of stamens which are inserted into its tube. Ovary superior, consolidated from two seed-leaves. Ovules ∞. Fruit a capsule or berry. Stipules 0.

### Family 1. OROBANCHEÆ.

Ovary 1-celled; placentæ parietal. Parasites without chlorophyll. Leaves reduced to scales.

1. **Anoplanthus** Endl. (*Aphyllon*).

Flowers without bractlets. Calyx 5-cleft. Corolla tubular, curved; limb 5-cleft. Stamens didynamous; cells of anthers separated, mucronate at their base. Placentæ 4. Capsule 1-celled, dehiscent by 2 valves; each valve with a placenta on each margin.—♃.

1. A. uniflorus L. Rhizome bearing only a few scapes; calyx lobes longer than the calyx-tube; flowers dull yellow, more or less tinged with violet; fragrant.—Saucelito. Rare. Spring.

2. A. FASCICULATUS Nutt. Rhizome bearing fasciculate scapes (peduncles); calyx lobes not longer than the calyx tube and much shorter than the corolla; flowers dull yellow, more or less tinged with purple.—Saucelito. Rare. Summer.

## 2. Aphyllon Mitchell.

Flowers with bractlets. Otherwise nearly as the preceding.—♃.

1. A. COMOSUM Gray. Paniculate-racemose; bractlets remote from calyx; calyx half the length of corolla; flowers pink.—Rare. Summer.

## Family 2. SCROPHULARIACEÆ.

Ovary 2-celled. Placentæ axillary or central. Capsule 2-celled, with 2 or 4 valves.

### Tribe 1. RHINANTHEÆ.

Lower half of corolla in æstivation covering the upper half. Corolla tubular; limb 2-labiate.

## 1. Pedicularis Tourn.

Didynamous. Corolla ringent; upper lip laterally compressed. Anthers of equal size and insertion. Capsule loculicidal.—♃.

1. P. DENSIFLORA Benth. Flowers crimson.— Hill-sides. Spring.

## 2. Cordylanthus Nutt.

Calyx spathaceous; fissure, if present, lateral; lips of corolla short; upper one laterally compressed. Stamens didynamous or diandrous. Anthers 2-celled; cells separated and of different shape and insertion. Capsule loculicidal.—☉.

1. C. MARITIMUS Nutt. Calyx 1-phyllous; flowers sessile, axillary; anthers of the two longer stamens 2-celled; anthers of the two shorter stamens dimidiate; leaves and bracts entire, pale green; corolla purplish.—Salt marshes, San Francisco. Summer.

2. C. MOLLIS Gray. Calyx 1-phyllous; flowers sessile, axillary, diandrous; anthers 2-celled; leaves and bracts linear, hirsute, the upper ones sometimes dentate, laciniate; flowers pale.—Salt marshes. Vallejo. Summer.

3. C. PILOSUS Gray. Didynamous. Calyx 2-phyllous; all anthers 2-celled; leaves and bracts soft, villous; flowers yellowish or purplish.—Santa Clara County, on dry grounds. Spring.

4. C. FILIFOLIUS Nutt. Didynamous. Calyx 2-phyllous, all anthers 2-celled; floral leaves hispid, flowers purplish.—Santa Cruz mountains. Spring.

### 3. Orthocarpus Nutt.

Calyx spathaceous, cleft vertically. Corolla personate, upper lip the smaller one, lower lip saccate. Didynamous; cells separated and of different shape and insertion. Capsule loculicidal.—☉.

1. O. LITHOSPERMOIDES Benth. Lower lip 3-saccate; sacs ample, ventricose, much larger than the upper lip; upper lip straight, slender, subulate, anther 2-celled; corolla white, tinged with purple.—Ocean Lake. Colma. Spring.

2. O. FAUCIBARBATUS Gray. Lower lip 3-saccate, much larger than the straight slender upper lip; anthers 1-celled; stamens wrapt in the involute upper lip; plant glabrous, only bracts puberulent; corolla pale.—Contra Costa. Spring.

3. O. ERIANTHUS Benth. Lower lip 3-saccate, much larger than the straight, slender upper lip; anthers 1-celled; stamens wrapt in the involute upper lip; plant pubescent; corolla yellow; in a variety (hybrid?) white or rose-color, but upper lip always dark.—Marin County. Spring.

4. O. FLORIBUNDUS Benth. Lower lip 3-saccate, much larger than the straight slender upper lip; anthers 1-celled; stamens not wrapt in the lanceolate upper lip; corolla white, tube twice the length of the calyx.—Marin County. Millbrae. Spring.

5. O. PUSILLUS Benth. Lower lip 3-saccate, larger than the slender, straight upper lip; anthers 1-celled; stamens not wrapt in the lanceolate upper lip; corolla purplish, its tube not surpassing the calyx.—Common. Spring.

6. O. PURPURASCENS Benth. Lower lip simply saccate, broader but not longer than the upper lip, the tip of which is incurved and back bearded; bracts and corolla white, rose or crimson.—Common. Spring.

7. O. CASTILLEIOIDES Benth. Lower lip ventricose; upper lip straight, its back not bearded; leaves lanceolate, commonly laciniate; upper leaves not attenuate, but cuneate-dilated and deeply cleft; spike dense; bracts and corolla pale, sometimes sprinkled with red.—Marsh near Tamalpais. Spring.

8. O. DENSIFLORUS Benth. Lower lip ventricose; upper lip straight, its back not bearded; leaves linear; upper leaves attenuate; bracts 3-cleft, as long as the flowers; spike dense; bracts and corolla purple and white.—Common. Spring.

9. O. ATTENUATUS Gray. Lower lip ventricose; upper lip straight, its back not bearded; leaves linear; bracts with slender lobes, herbaceous and barely white-tipped; flowers pale; lower lip purple-spotted; spike virgate.—Common. Spring.

### 4. Castilleia Nutt. PAINTED-CUP.

Calyx spathaceous, cleft vertically. Corolla ringent, upper lip the larger. Stamens didynamous. Cells of anthers separated and of different shape and insertion. Capsule loculicidal.

1. C. FOLIOLOSA Hook. & Arn. Dorsal and ventral side of calyx equally cleft; plant tomentose; leaves linear; bracts crimson.—♃. Common. Summer.

2. C. PARVIFLORA Bong. Dorsal and ventral side of calyx equally cleft; plant villous, viscid; leaves laciniate, lobes linear; bracts crimson, yellow or white.—♃. San Mateo. Summer.

3 C. LATIFOLIA Hook. & Arn. Dorsal and ventral side of calyx equally cleft; plant villous, viscid; leaves oval, obtuse, usually entire; when lobed, lobe obtuse, oval; bracts crimson, yellow or white.—♃. Common. Summer.

4. C. AFFINIS Hook & Arn. Ventral side of calyx more deeply cleft than dorsal side. Calyx and upper part of the bracts petaloid, crimson, sometimes orange.—♃. Common. Summer.

### Tribe II. VERONICEÆ.

Lower half of the corolla in æstivation covering the upper half. Corolla not tubular nor 2-labiate.

### 1. Veronica L.  SPEEDWELL. BROOKLIME.

Calyx 4-5-cleft. Corolla rotate, 4-lobed, upper lobe the largest. Ovary few-seeded. Capsule compressed, emarginate.

1. V. PEREGRINA L. Upper leaves alternate, sessile; lowest leaves opposite, petiolate; flowers single, axillary bluish.—☉. Low grounds. Spring.

2. V. AMERICANA Schweinitz. All leaves opposite; flowers in axillary racemes; corolla bluish.—♃. Aquatic. Not common. Summer.

### 2. Synthyris Benth. (*Wulfenia* Jacq.)

Calyx 4-cleft. Corolla campanulate. Limb of corolla 4-lobed; posterior lobe emarginate, diandrous. Capsule 2-celled. Ovules ∞.

1. S. ROTUNDIFOLIA Gray. Flowers purplish.—♃. Lagunitas Creek. Spring.

### 3. Limosella L.  MUDWORT.

Calyx 5-dentate. Didynamous. Limb of corolla 5-cleft. Cells of anthers confluent, transversely dehiscent. Capsule 1-celled, ∞-seeded.—☉.

1. L. AQUATICA L. Flowers pale.—Salt marshes. Summer.

### Tribe III.  GRATIOLEÆ.

Upper lip in æstivation covering the lower. Stigma flat, 2-lobed.

### 1. Gratiola L.

Calyx 5-parted, with two bractlets. Corolla 2-labiate. Filaments 4; two of them fertile, two sterile (sometimes 0). Ovary ∞-ovulate. Capsule loculicidal, at length septifragal.

1. G. EBRACTEATA Benth. Corolla pale. ⊙.—Marin County. Spring.

### 2. Mimetanthe Greene.

Calyx campanulate, not angulate. Capsule dehiscent only by the dorsal suture.

1. M. PILOSA Greene. Corolla yellow.—⊙. Dry river beds. Summer. The whole plant has a nauseous smell, resembling that of *Datura stramonium*.

### 3. Mimulus L. MONKEY-FLOWER.

Calyx angular; angles carinate. Placentæ of the capsule remain united and only separate at last near the apex. Valves of the capsule membranaceous.

1. M. INCONSPICUUS Gray. Glabrous; leaves ovate, entire, 3-5 nerved; all cauline leaves sessile; calyx ventricose, its teeth about equal and very short; corolla yellow.—⊙. Livermore. Summer.

2. M. NASUTUS Greene. Glabrous; stem 4-angular; leaves dentate, 3-5 nerved, mostly petioled; calyx ventricose, in fruit conspicuously pointed by the projecting upper tooth;

corolla yellow, commonly with a purple spot on the lower lip.—⊙ Common in marshy places. Summer.

3. M. LUTEUS L. Glabrous; leaves variable in form, dentate; upper leaves sessile, lower leaves petiolate; calyx ventricose, inflated; teeth unequal, but in fruit without the projection of the upper largest tooth; corolla yellow, marked with purple.—♃. Common in marshy places. Summer.

4. M. MOSCHATUS Dougl. Villous and viscous; leaves pinnately veined; corolla yellow.—♃. Wet places in the mountains. Summer.

5. M. CARDINALIS Dougl. Villous and viscous; leaves ovate, the upper connate; limb of corolla very oblique; corolla scarlet.—♃. Water-courses. Contra Costa. Summer.

### 4. Diplacus Nutt. STICKY MONKEY-FLOWER.

Calyx angular; angles carinate. Placentæ meeting, but in dehiscence separate through their whole length. Valves of the capsule coriaceous. Tube of corolla funnel-shaped.

1. D. GLUTINOSUS Wendt. Flowers orange-color, but running sometimes into different shades of yellow and red.— ♄. Common. Summer.

### 5. Eunanus Benth.

Calyx angular; angles carinate. Placentæ of the capsule separate in dehiscence for their whole length. Tube of corolla slender, filiform.

1. E. Douglasii Benth. C o r o l l a red, sometimes s p o t t e d .—☉. Marin County. Spring.

### Tribe IV. Digitaleæ.

Upper lip of corolla in æstivation covering the lower lip. Stigma minute, not flat. Capsule septicidal.

#### 1. Pentstemon L'Herit.

Didynamous, with a fifth sterile filament. Calyx 5-parted. Limb of corolla labiate.—♃.

1. P. azureus Benth. Anthers sagittate; their cells confluent at the apex, dehiscent by a common split, that does not extend beyond the middle of the cell; corolla blue. Niles. Summer.

2. P. centranthifolius Benth. Cells of the anthers divaricate; dehiscent for their whole length; corolla scarlet. Niles. Summer.

### Tribe V. Antirrhineæ.

Upper lip in æstivation covering the lower lip. Tube of corolla gibbous or calcarate. Capsule neither loculicidal nor septicidal.

## 1. Collinsia Nutt.

Calyx 5-parted. Corolla personate, upper lip erect, lower lip 3-lobed, middle lobe laterally compressed, hiding the didynamous stamens. Capsule few-seeded, septifragal.—☉.

1. C. PARVIFLORA Dougl. Flowers with long pedicels, mostly solitary in the axils of the leaves; throat of corolla longer than the limb; corolla small, blue; upper lip sometimes white. Saucelito. Spring.

2. C. SPARSIFLORA Fisch & Meyer. Flowers with long pedicels; mostly solitary in the axils of leaves; throat of corolla shorter than the limb; corolla pale-purple lower lip violet.—Marin County. Ocean Lake. Spring.

3. C. BARTSIÆFOLIA Benth. Flowers with short pedicels, crowded in the axils of leaves or bracts; throat of corolla longer than broad; corolla pale-p u r p l i s h. — Mission Dolores. Spring.

4. C. BICOLOR Benth. Flowers with short pedicels, crowded in the axils of leaves or bracts; throat of corolla as broad as long, inflated and saccate; upper lip of corolla pale, lower lip violet.—San Rafael. Spring.

## 2. Antirrhinum L. SNAPDRAGON.

Didynamous. Calyx 5-cleft. Corolla saccate, 2-labiate, personate. Capsule dehiscent by pores at the apex.

1. A. VAGANS Gray. Corolla violet.—☉.
Contra Costa. San Mateo. Spring.

### 3. Linaria Tourn. TOAD-FLAX.

Didynamous. Calyx 5-cleft. Corolla with a spur, 2-labiate, personate. Capsule separating at the apex into two valves..

1. L. CANADENSIS Dumont. Corolla blue.—☉. Common. Spring.

### Tribe VI. VERBASCEÆ.

Upper lip in æstivation covering the lower lip. Corolla neither ringent nor personate. Capsule septicidal.

### 1. Scrophularia Tourn. FIGWORT.

Corolla short, globular. Limb narrow, 5-lobed. Middle lobe of lower lip reflexed. Didynamous (rudiment of a fifth stamen).

1. S. CALIFORNICA Cham. Corolla dark purple.—♃. Common. Summer.

Order 5. POLEMONIALES. Stamens as many as the lobes of the corolla, and more than the cells of the ovary. Stipules 0.

### Family 1. SOLANACEÆ.

Corolla regular. Ovary 2–4 celled; ∞-ovulate. Fruit capsule or berry.

### 1. Solanum L. POTATO. NIGHTSHADE.

Calyx persistent in fruit. Corolla rotate; in æstivation valvate, induplicate. Anthers connivent, opening by apical pores. Fruit an ∞-seeded berry.

1. S. NIGRUM L. Corolla deeply 5-cleft; white or blue; berry black or orange.—☉. Common throughout the year.

Very variable species. The berries of most, perhaps of all our Californian variations are not poisonous, and are frequently eaten by children.

2. S. UMBELLIFERUM Esch. Corolla merely 5-angled, rotate, blue; berries red.—♃. ♄. Common. Flowering at all seasons. The berries of this species are at least suspicious.

### 2. Datura L. THORN-APPLE.

Calyx tubular, circumscissile. Corolla funnel-shaped, plicate and convolute in æstivation. Anthers dehiscent longitudinally. Ovary spuriously 4-celled. Capsule ovate, septifragal. Seeds reniform.

1. D. STRAMONIUM L. Corolla white, tinged with pale blue.—☉. Sunnyside. Summer.

Poisonous ballast-weed; differs from the European and Asiatic type by the bluish tinge of its corolla.

### 3. Nicotiana L. TOBACCO.

Calyx tubular, peristent. Corolla funnel or salver-shaped. Anthers longitudinally dehiscent. Ovary 2-celled, ∞-ovulate. Capsule septicidal. Seeds ∞, minute.

1. N. Bigelovii Watson. Flowers white. ⊙. Buena Vista, Sonoma. Summer.

## Family 2. BORRAGINEÆ.

Ovary 4-celled, 4-lobed. Cells 1-seeded. Fruit 4 nutlets. Style 1.

### 1. Amsinckia Lehm.

Calyx 5-parted. Corolla salver- or funnel-shaped. Limb 5-lobed. Ovary 4-lobed. Style central. Fruit 4 nutlets, attached to a conical disc above its base.—⊙. Inflorescence circinate. Flowers yellow.

1. A. vernicosa Hook & Arn. Nutlets triquetrous, straight, smooth, shining; attached at the lower part of the inner angle.—Contra Costa. Spring.

2. A. tesselata Gray. Nutlets testaceous, tessellate, rugose; calyx lobes obtuse; corolla orange-color.—Contra Costa. Spring.

3. A. intermedia Fisch. & Mey. Nutlets scabrous, convex and carinate on the back; calyx-lobes narrow, acute; corolla tube scarcely exceeding the calyx.—Paper Mill Creek. Spring.

4. A. spectabilis Fisch. & Mey. Nutlets granulate, convex and carinate on the back; calyx lobes linear; corolla orange, tube at least twice the length of the calyx lobes.—Contra Costa. Spring.

6. A. lycopsoides Lehm. Nutlets reticulate, rugose; calyx lobes lanceolate.—Common. Spring.

### 2. **Krynitzkia** Fisch. & Mey.
(*Eritrichium* Schrad).

Nutlets ventrally attached to the conical or columnar disc from near their base.

1. K. muriculata Gray. Nutlets attached to the disc from the base nearly up to the apex; calyx in fruit somewhat longer than the nutlets, its sepals lanceolate, armed with pungent bristles; mid-rib of sepals of usual shape; nutlets muricate; corolla white. — ⊙. Common. Spring.

2. K. ambigua Gray. Nutlets attached to the disc from the base up to nearly the apex; calyx in fruit considerably longer than the nutlets, its sepals narrow, armed with pungent bristles; mid-rib of sepals prominent; nutlets scabrous; corolla white. — ⊙. Common. Spring.

3. K. Torreyana Gray. Nutlets attached to the pyramidal disc only up to their middle; nutlets smooth, all four maturing; corolla white.—⊙. Saucelito. Spring.

4. K. leiocarpa Fisch. & Mey. Nutlets attached to the subulate disc nearly their whole length, smooth, all four maturing; corolla white.—⊙. Saucelito. Spring.

5. **K. oxycarya** Gray. Nutlets attached only by their lower third to the conical gynobase; only one of the four maturing; corolla white.—⊙. Marin County. Spring.

6. **K. chorisiana** DC. Nutlets attached to the disc only by their base; corolla nearly rotate; scales near the throat conspicuous, puberulent; corolla white.—⊙. Wet ground, San Francisco. Spring.

7. **K. californica** DC. Nutlets attached to the disc only by their base; corolla inconspicuous; scales near the throat not puberulent; corolla white.—⊙. Wet grounds. Common. Spring.

### 3. Plagiobothrys Fisch. & Mey.
(*Eritrichium* Schrad).

Nutlets attached near the middle of their ventral face to the conical disc.

1. **P. nothofulvus** Gray. Calyx cleft only to its middle; corolla white.—⊙. Common. Spring.

2. **P. canescens** Benth. Calyx cleft nearly to the base; corolla white.—⊙. Livermore. Spring.

### 4. Pectocarya DC.

Corolla funnel-shaped; throat closed. Style central, very short. Fruit four nutlets, marginate and in pairs,—⊙. Flowers axillary, sessile, white.

1. P. penicillata A. DC.—Marin County. Spring.

### 5. Cynoglossum L.    Hound's-tongue.

Corolla funnel-shaped. Throat closed by five scales. Ovary 4-lobed. Style central. Fruit 4, depressed, echinate nutlets.—♃. Inflorescence paniculate, ebracteate, circinate.

1. C. grande Dougl. Corolla blue.—Common in woods. Spring.

### 6. Heliotropium L.    Heliotrope.

Corolla salver-shaped. Limb 5-lobed; sinuses plicate. Ovary 4-celled. Cells 1-seeded. Style terminal, short. Stigma peltate. Fruit a 4-pyrenous drupe. Inflorescence circinate.

1. H. curassavicum L. Corolla pale.—♃. Seashore. Summer.

### Family 3. HYDROPHYLLEÆ.

Ovary incompletely 2-celled. Style 2-cleft or 2 styles. Fruit a capsule.

### 1. Eriodictyon Benth.    Yerba Santa.

Corolla campanulate. Ovary 2-celled; cells ∞-ovulate: styles 2; stigmas clavate. Fruit a capsule, 2-celled, few-seeded, at first loculicidal, at last septicidal.—♄. Inflorescence circinate, arranged in a panicle. Flowers blue or white.

1. E. GLUTINOSUM Benth. Corolla pale or bluish.—Chaparral and chemisal. Summer.

An exceedingly variable species. The resinous exudation of the young branches and leaves has a bitter, somewhat aromatic taste. The plant is called "Yerba Santa." It acts as a mild diuretic, and was used as such by the old inhabitants of California. At present considerable quantities are exported, partly for medicinal purposes, partly as a harmless and agreeable substitute for hops in brewing certain varieties of beer, especially porter.

## 2. Romanzoffia Cham.

Corolla salver-shaped. Style entire. Stigma capitate. Ovary incompletely 2-celled, $\infty$-ovulate. Capsule ovate, loculicidal. Seeds $\infty$, very small.—♃. Inflorescence loosely circinate.

1. R. SITCHENSIS Bongard. Corolla pale.—Sunnyside. Paper Mill Creek. Spring.

## 3. Emmenanthe Benth.

Calyx 5-parted. Sinuses naked; lobes equal. Corolla campanulate, persistent. Ovary incompletely 2-celled, $\infty$-ovulate. Stigma 2-cleft.—☉. Inflorescence circinate. Flowers yellow.

1. E. PENDULIFLORA Benth. Corolla cream-colored.—Livermore. Summer.

## 4. Phacelia Juss.

Calyx 5-parted. Sinuses naked. Lobes equal. Corolla campanulate, deciduous. Ovary incompletely 2-celled or 1-celled, with two parietal placentæ. Number of ovules variable. Stigma 2-cleft. Capsule incompletely 2-celled or 1-celled, loculicidal. Inflorescence circinate, cymose.

1. P. CIRCINATA Jacq. Ovary contains four ovules, inflorescence thyrsoid, circinate, with very short pedicels; leaves entire, the lowermost tapering into a petiole with one or two pairs of leaflets; corolla blue or white.—♃. Common. Summer.

2. P. MALVÆFOLIA Cham. Ovary contains four ovules; circinate spikes solitary or geminate, not collected into a thyrsoid inflorescence; all leaves petiolate, cordate, lobed, dentate; corolla pale.—⊙. San Francisco. Summer. Leaves sting somewhat like nettles.

3. P. HISPIDA Gray. Ovary contains four ovules; inflorescence thyrsoid, circinate; flowers with short pedicels; sepals very narrow, much longer than the capsule; leaves about 5-parted; corolla red.—⊙. Livermore. Spring.

4. P. DISTANS Benth. Ovary contains four ovules; inflorescence thyrsoid, circinate; flowers sessile; sepals unequal and much longer than the capsule; leaves finely and decom-

poundly dissected; stamens scarcely exserted; corolla whitish, ochroleucous or violet.—☉. Alameda. Marin County. Summer.

5. P. TANACETIFOLIA Benth. Ovary containing four ovules; inflorescence thyrsoid, circinate; flowers with very short pedicels; sepals but little longer than the capsule; leaves finely and decompoundly dissected; stamens much exserted; corolla bluish.—☉. Strawberry Valley. Contra Costa. Summer.

6. P. CILIATA Benth. Ovary containing four ovules; inflorescence cymose, circinate; the circinna rather short; sepals in fruit accrescent and transversely veined; leaves pinnately parted, the partitions pinnatifid; stamens somewhat shorter than the corolla; white or blue. ☉.—Belmont. Spring.

7. P. DIVARICATA Gray. Ovary containing more than four ovules; inflorescence in racemes; pedicels short; leaves ovate, entire; corolla blue.—☉. Tiburon. Spring;

### 5. Ellisia L.

Calyx 5-parted, sinuses naked. Corolla campanulate, stamens included. Ovary 1-celled with two parietal placentæ. Stigma 2-cleft. Capsule membranaceous, loculicidal, the placentæ separating from the capsular valves, simulating a second internal capsule. ☉.—Flowers pale.

1. E. CHRYSANTHEMIFOLIA Benth. Corolla whitish.—Oakland hills. Spring.

### 6. Nemophila Nutt.

Calyx with reflexed, appendiculate sinuses. Corolla campanulate. Ovary 1-celled, with two parietal placentæ. Stigma 2-cleft. Capsule membranaceous, loculicidal, the placentæ separating from the capsular valves, simulating a second internal capsule.—☉.

1. N. INSIGNIS Dougl. Ovary containing more than four ovules; scales on the base of corolla rounded and partly free; corolla blue. Common. Spring.

2. N. MENZIESII Hook. & Arn. Ovary containing more than four ovules; scales at the base of corolla narrow and wholly adnate; corolla blue or white, more or less marked by dots.—Common. Spring.

3. N. AURITA Lindl. Ovary containing but four ovules; scales on the base of corolla in pairs, broad and partly free; corolla violet. San Mateo. Spring,

4. N. PARVIFLORA Dougl. Ovary contains but 4 ovules; scales on the base of corolla oblong, wholly adnate by one edge; corolla light blue or white.—Common in shady places. Spring.

### Family 4. POLEMONIACEÆ.

Æstivation imbricate. Ovary 3-celled. Style 3-cleft. Embryo straight. Cotyledons foliaceous.

## 1. Polemonium Tourn. JACOB'S-LADDER.

Corolla rotate. Stamens ascending. Filaments dilated at their base. Seeds $\infty$.

1. P. CARNEUM Gray. Corolla blue.—♃. Point Bonita, near the lighthouse. Summer.

## 2. Gillia Ruiz & Pavon.

Corolla funnel or salver-shaped. Stamens straight, equally inserted.

1. G. PUSILLA Benth. Leaves opposite, uppermost sometimes alternate, palmately divided; divisions filiform; corolla short, funnel-shaped; pedicels capillary; corolla purplish, with yellowish throat, sometimes pale.—☉. Marin County. Spring.

2. G. DICHOTOMA Benth. Leaves all opposite; corolla salver-shaped, its tube shorter than the calyx; flowers almost sessile, but not capitate; corolla white.—☉. Livermore. Spring.

4. G. DENSIFLORA Benth. Leaves opposite, palmately parted, fascicled in the axils; corolla salver-shaped, its tube about the length of the obovate lobes; flowers in glomerules; bracts herbaceous; corolla red or white.—☉. Alameda. San Rafael. Spring.

4. G. ANDROSCAEA Steud. Leaves opposite, palmately parted; corolla salver-shaped, its tube more than twice the length of the lobes;

flowers in glomerules; bracts herbaceous, hirsute, much shorter than the corolla tube; corolla red or white, with yellowish throat.—☉. Common. Spring.

5. G. MICRANTHA Steud. Leaves opposite, palmately parted; corolla salver-shaped, its tube more than four times the length of the lobes; flowers in glomerules; bracts herbaceous, pubescent, shorter than the flower-tube; corolla red, white, lilac or yellow.—☉.

6. G. TENELLA Benth. Leaves opposite, palmately parted; corolla salver-shaped, its tube more than four times the length of the corolla lobes; flowers in glomerules; bracts herbaceous, hispidulous, shorter than the corolla tube; corolla purple or pink, with yellow throat.—☉. Common. Spring.

7. G. CILIATA Benth. Leaves opposite, palmately parted; corolla salver-shaped, its tube more than four times the length of the lobes; flowers in glomerules; bracts very hirsute, about the length of the corolla tube; calyx lobes acerose; corolla rose, violet or pale.—☉. Marin County. Spring.

8. G. SQUARROSA Hook. & Arn. Leaves alternate, pinnately parted, partitions parted or incised; corolla funnel-shaped; flowers in glomerules, densely bracteate; bracts rigid, pungent; stamens included in the tube of the corolla; corolla blue.—☉. Common. Summer.

The disagreeable smell of this plant is the cause of its California name, "Skunkweed."

9. G. COTULÆFOLIA Steud. Leaves alternate, 2-pinnately parted; corolla funnel-shaped; flowers in glomerules, densely bracteate; bracts spinescent; stamens exserted; ovules no more than two in each cell; corolla bluish or pale.— ⊙. Common. Summer.

Smell somewhat like Chamomile.

10. G. INTERTEXTA Steud. Leaves alternate, pinnately parted, partitions but little, if at all, divided; flowers in glomerules, densely bracteate; bracts with very villous base; corolla funnel-shaped, not exceeding the calyx lobes; stamens exserted; ovules more than two in each cell; corolla small and white.—⊙. Common. Summer.

11. G. LEUCOCEPHALA Gray. Leaves alternate, pinnately parted, soft; partitions slender and frequently entire; flowers in glomerules; bracts barely pungent; corolla funnel-shaped, exceeding the calyx-lobes; stamens considerably exserted: ovules two in each cell; corolla white.—⊙. Common. Summer.

12. G. VISCIDULA Gray. Leaves alternate, pinnately parted; partitions entire, subulate: flowers in glomerules; bracts dilated at their base; corolla funnel-shaped, about twice the length of the spinescent calyx-lobes; plant

viscid, pubescent; corolla violet.—☉. Tamalpais. Summer.

13. G. DENSIFOLIA Benth. Leaves alternate, pinnately laciniate, lobes spinulose; corolla salver-shaped; stamens exserted; flowers in glomerules; bracts foliaceous; stems leafy to their tops; corolla blue.—♃. Livermore. Summer.

14. G. VIRGATA Steud. Leaves alternate, filiform, simple or 3-parted, the partitions filiform; corolla salver-shaped; stamens exserted; flowers in glomerules; bracts foliaceous; stems with but few leaves; virgate; corolla blue.—☉. Livermore. Summer.

15. G. CAPITATA Dougl. Leaves alternate, 2-3-pinnately divided into slender lobes; corolla funnel-shaped, its throat being very slightly dilated; flowers crowded but not sessile, the cluster itself on a long stalk; bracts inconspicuous; calyx glabrous; stamens inserted into the sinuses of the light blue corolla. ☉.—Marin County. Summer.

16. G. ACHILLEÆFOLIA Benth. Leaves alternate, 2-3-pinnately divided into slender lobes; corolla somewhat funnel-shaped, but its throat abruptly and amply dilated; flowers in clusters but not sessile; bracts inconspicuous; calyx woolly; corolla blue.—☉. Sand hills. San Francisco. Summer.

17. G. MULTICAULIS Benth. Leaves alternate, 2-pinnately parted into linear lobes; corolla funnel-shaped, its tube shorter than the calyx and about the length of the ovate corolla lobes; flowers in clusters, pedicels in fruit about equaling the calyx; bracts inconspicuous; corolla violet.—⊙. Oakland hills. Summer.

18. G. TRICOLOR Benth. Leaves alternate, 2-pinnately parted into linear lobes; corolla much longer than the calyx with very short tube, but ample, funnel-shaped throat; lobes of the corolla longer that the stamens; flowers in short peduncled clusters; pedicels very short; bracts inconspicuous; tube of the corolla yellowish, throat marked with purple; lobes violet.—⊙. Livermore. Summer.

19. G. INCONSPICUA Dougl. Leaves alternate, pinnatifid; lobes short, mucronate; corolla narrowly funnel-shaped, about the length of the calyx; bracts inconspicuous; corolla violet.—⊙. Common. Summer.

### 3. Collomia Nutt.

Corolla salver-shaped, with long tube. Stamens unequally inserted.

1. C. GRACILIS Dougl. All leaves sessile, entire; corolla violet; the tube yellowish.—⊙. Tamalpais, Spring.

The seeds when moistened develop a mucilage, which in Mexico is called "Chia" and used in the preparation of a cooling drink.

2. C. GILIOIDES Benth. Lower leaves petiolate and most of them pinnately incised; calyx rounded at base; corolla pink.—☉. Tamalpais. Spring.

3. C. HETEROPHYLLA Hook. Lower leaves petiolate, 2-pinnatifid; calyx acute at its base; corolla pink.—☉. Lagunitas Creek. Spring.

### Family 5. CONVOLVULACEÆ.

Æstivation contorted. Ovary 2-3 or 4-celled. Embryo curved. Cotyledons foliaceous, conduplicate, corrugate.

#### 1. Dichondra Forst.

Ovary divided into two carpidia, each 2-ovulate. Creeping herb. Flowers inconspicuous.

1. D. REPENS Forst. Flowers pale.—♃. Telegraph hill. Spring.

#### 2. Convolvulus L.
BIND-WEED. MORNING-GLORY.

Calyx 5-cleft, persistent. Corolla funnel-form, 5-angulate, 5-plicate. Cells of the capsule 2-ovulate.—♃.

1. C. SOLDANELLA L. Margin of corolla entire; stigmas ovate, thickish; corolla pink. Seashore. Summer.

2. C. CALIFORNICUS Choisy. Margin of corolla entire; stigmas linear-oblong, flat; bracts at the base of the calyx; corolla white or reddish.—Berkeley. Summer.

3. C. LUTEOLUS Gray. Margin of corolla entire; stigmas linear, flat; bracts distant from the calyx; corolla pale-yellow, sometimes reddish.—Common. Summer.

4. C. ARVENSIS L. Margin of corolla entire; stigmas filiform; bracts at the base of the pedicel and minute; corolla white, sometimes tinged with rose.—Common. Summer.

5. C. PENTAPETALOIDES L. Corolla deeply 5-cleft, purplish.—☉. Not common. Contra Costa. Summer.

### 3. Cressa L.

Corolla funnel-shaped, 5-parted, not plicate. Styles 2. Cells of the ovary 2-ovulate. Fruit a 1-seeded capsule.—♃.

1. C. CRETICA L. Corolla white, silky outside.—Seashore. Summer.

### 4. Cuscuta Tourn. DODDER.

Aphyllous. Destitute of chlorophyll. Corolla urceolate. Limb 4-5-lobed. Ovary 2-celled; cells 2-ovulate. Capsule indehiscent, sometimes circumscissile. Embryo spiral. Cotyledons 0.

1. C. SALINA Engelm. Calyx lobes as long as the tube of the white corolla.—Common. Summer. Parasite of *Salicornia* and other salt-marsh herbs.

2. C. SUBINCLUSA Durand & Hilgard. Calyx lobes much shorter than the tube of corolla; corolla white, calyx usually reddish.—Parasites on *Ceanothus* and other shrubs.—Not common. Summer.

Section II. ISOCARPICÆ (Number of carpidia corresponding to number of floral parts).

### Order 1. PRIMULALES.

Stamens opposite to the lobes of corolla. Ovary 1-celled. Placenta central.

#### Family 1. LENTIBULARIÆ.

Diandrous. Corolla irregular, spurred.

1. **Utricularia** L.  BLADDERWORT.

1. U. VULGARIS L. Corolla yellow. Aquatic.—Olema. Summer.

#### Family 2. PLUMBAGINEÆ.

All parts in fives except the ovule, which is single.

1. **Armeria** Willd.  THRIFT.

Flowers in an involucrate head. Corolla 5-parted or five distinct petals. Styles 5.—♃.

1. A. VULGARIS Willd. Corolla rose-color. Common. Spring.

## 2. Statice L.   Sea-Lavender.

Flowers bracteate in one-sided spikes.—♃.

1. S. Limonium L.   Corolla violet.—Salt marshes. Summer.

### Family 3. PRIMULACEÆ.

Pentandrous. Placenta central, ∞-ovulate.

#### 1. Dodecatheon L.   Shooting Star.

Flowers umbellate. Filaments shorter than the anthers, connivent in a cone. Acaulescent. Flowers purple, pink or white.—♃.

2. D. meadia L.—Common. Spring.

#### 2. Androsace Tourn.

Flowers umbellate. Filaments shorter than the anthers, included in the salver-shaped corolla.

1. A. occidentalis Pursh. Flowers white.— ⊙. Contra Costa Range. Santa Cruz mountains. Spring.

#### 3. Glaux L.   Milkwort.

Calyx campanulate, 5-cleft, colored. Corolla 0. Stamens 5, alternate with the lobes of the calyx. Capsule 5-valved, few-seeded.—♃. Leaves decuscate; flowers white.

1. G. maritima L.—Salt marshes. Summer.

#### 4. Trientalis L.   Star-flower.

Flowers several, terminal. Floral parts 5 to 7. Capsule few-seeded, longitudinally dehiscent.—♃.

1. LATIFOLIA Torrey. Corolla pink or pale.— Shady woods. Spring.

### 5. Anagallis Tourn. PIMPERNEL.

Flowers axillary. Lobes of the corolla broad. Capsule transversely dehiscent. Leaves opposite.

1. ARVENSIS L. Corolla red, sometimes pale, purple or blue. — ☉. Common all the year round. An infusion of this herb was used by the old Californians for headache.

### 6. Samolus Tourn. BROOKWEED.

Calyx half-superior. Corolla campanulate. Stamens 10; the five fertile opposite to the lobes of the corolla; the five sterile, alternate. Capsule 5-valved, ∞-seeded. Leaves alternate; flowers white.—♃.

1. L. VALERANDI L. Rare. Formerly in a marsh near San Francisco; at present found occasionally in Marin County.

ORDER 2. ERICALES. Stamens alternate with the lobes of corolla or twice as many. Placentæ axillary.

### Family 1. PYROLACEÆ.

Corolla split into petals. Hypogynous disc 0. Seeds minute.

### 1. Pyrola Tourn.

Calyx 5-parted. Corolla 5-petalous. Stamens 10. Style filiform.—♃. In forest shades.

1. P. APHYLLA Smith. Parasitic without chlorophyll; petals white.—Lagunitas Creek. Summer.

### Family 2. RHODORACEÆ.

Ovary free. Fruit a capsule, septicidal.

### 1. Rhododendron L.

Corolla funnel-shaped, 5-lobed. Stamens ascending. Cells of anthers dehiscent by an apical pore. Seeds ∞, minute.— ♄.

1. R. OCCIDENTALE Gray. Corolla white, upper lobes yellow.—Point Bonita. Taylorville. Sonoma. Summer. (Azalea.) The root of this shrub contains a powerful narcotic.

### Family 3. ERICACEÆ.

Gamopetalous. Fruit either baccate or a loculicidal capsule.

### 1. Gaultheria L.   WINTERGREEN.

Calyx 5-cleft, petaloid. Corolla urceolate. Stamens 10. Cells of anthers dehiscent by an apical pore. Fruit a spurious berry, i. e., a capsule, 5-celled, ∞-seeded, enclosed by the enlarged and fleshy calyx.—Leaves evergreen. Flowers white, rosy.— ♄.

1. G. Shallon Pursh. Berry red, at last black. Edible. Saucelito. Tamalpais.

### 2. Arctostaphylos Adans. Manzanita.

Corolla urceolate. Stamens 10. Disc hypogynous. Cells of ovary 1-ovulate. Fruit a berry. Cells 1-seeded; seeds sometimes coherent by their covering. Leaves evergreen. Flowers white, rosy.— ♄. Fruit edible.

1. A. pungens HBK. Almost glabrous; leaves entire, coriaceous, mucronate; petioles slender; pedicels glabrous; drupe glabrous; corolla white. — ♄. Common. Spring.

2. A. tomentosa Dougl. Young branchlets tomentose, old ones bristly; leaves almost entire, coriaceous, petioled; ovary hirsute; drupe puberulent, but becoming glabrous at last; corolla white.—Hillsides. Spring.

3. A. Andersonii Gray. Branchlets puberulent and bristly; leaves generally conspicuously serrulate, thin, almost sessile; drupes depressed, densely covered with viscous bristles. Livermore. Spring.

### 3. Arbutus Tourn. Madroña.

Corolla urceolate. Stamens 10. Disc hypogynous. Cells of ovary $\infty$-ovulate. Fruit a berry. Cells several-seeded. Leaves evergreen. Flowers white.— ♄.

1. A. Menziesii Pursh.—Common. Spring. Fruit edible.

### Family 4. VACCINIEÆ.

Ovary inferior.

#### 1. Vaccinium L   Huckleberry.

Anther cells separate, elongated at the apex into a tube. Fruit a berry, ∞-seeded, crowned by the persistent calyx.

1. V. ovatum Pursh.   Corolla pink.— ♄. Redwoods. Spring.

### Sub-Series 2. Polypetalæ.

Section 1. Discophoræ. Ovary inferior and crowned by a well developed disc.

Order 1. UMBELLALES. Cells of ovary 1-ovulate. Stamens alternate.

### Family 1. UMBELLIFERÆ.

Stamens 5. Ovary 2-celled. Styles 2. Fruit a schizocarp. (Cremocarp.)

#### 1. Caucalis L.

Umbels regularly compound. Secondary ribs of the cremocarp more prominent than the primary. Margins of the endosperm inflexed ⊙.—Flowers white.

1. C. nodosa Hudson. Decumbent; umbels naked.—Marin County. Livermore. Spring.

2. C. microcarpa Hook. & Arn. Erect; umbels involucrate.—Common. Spring.

The chewing of this herb, called by the Spanish "Yerba de vibora," or an infusion of it, is recommended by them as an antidote against the bite of the rattlesnake.

### 2. Daucus Tourn. Carrot.

Umbels regularly compound. Secondary ribs of the cremocarp more prominent than the primary. Intervals 1-vittate. Face of the endosperm flat.

1. D. pusillus Michx. Flowers white.—☉. Common. Spring.

### 3. Ferula Tourn.

Umbels regularly compound. Cremocarp dorsally compressed. Marginal ribs winged. Wings coherent. Dorsal ribs filiform. Vittæ ∞. Flowers yellow.—♃.

1. F. Californica Gray.—Sunnyside. Summer.

### 4. Heracleum L. Cow-Parsnip.

Umbels regularly compound. Cremocarp dorsally compressed. Marginal ribs winged. Wings coherent. Vittæ shorter than the mericarp.—♃. Flowers white.

1. H. lanatum Michx.—Common. Summer. Herb of suspicious qualities.

## 5. Peucedanum L.

Umbels regularly compound. Cremocarp dorsally compressed. Marginal ribs winged. Wings coherent. Vittæ as long as the mericarp.—♃.

1. P. DASYCARPUM Torr. & Gray. Leaves much dissected; cremocarp tomentose; flowers white.—Common. Summer.

Smell of entire plant like celery.

2. P. UTRICULATUM Nutt. Leaves much dissected; cremocarp glabrous, distinctly ribbed; flowers yellow.—Common. Summer.

3. P. CARUIFOLIUM Torr & Gray. Leaves much dissected; cremocarp glabrous; dorsal ribs obsolete, only the marginal ones developed; flowers yellow.—Common. Summer.

4. P. TRITERNATUM Nutt. Leaves 2-ternate to 3-quinate; umbellules involucellate; flowers yellow.—Contra Costa. Summer.

5. P. LEIOCARPUM Nutt. Leaves 2-ternate to 3-quinate; umbellules naked; flowers yellow.—Livermore. Summer.

The roots of several species of *Peucedanum* have been used as food, but as those used in this way are not sufficiently identified it is safer to abstain from using them till the different species have been subjected to an examination, not merely in regard to their botanical characters, but also in regard to their effects on the human organism.

### 6. Angelica L.

Umbels regularly compound. Cremocarp dorsally compressed. Marginal ribs winged. Wings distinct; intervals 1-vittate. Flowers white.—♃.

1. A. TOMENTOSA Watson.—Common. Summer. Properties like those of the officinal *Angelica Archangelica*.

### 7. Selinum L.

Umbels regularly compound. Cremocarp dorsally compressed, 4-winged by very distinct marginal ribs, and only coherent by a carinate commissure.—♃. Flowers white.

1. S. PACIFICUM Watson.—Saucelito. Summer.

### 8. Ligusticum L.  LOVAGE.

Umbels regularly compound. Cremocarp dorsally compressed, 4-winged by the very distinct lateral ribs. Mericarps coherent by a convex commissure.—♃.

1. L. APIIFOLIUM Benth. & Hook. Flowers white.—Tamalpais. Summer.

### 9. Œnanthe L.  FOOL'S PARSLEY.

Umbels regularly compound. Cremocarp terete, ovate. Ribs obtuse. Intervals 1-vittate. Aquatic. Flowers white.—♃.

1. Œ. CALIFORNICA Watson.—Common. Aquatic. Summer.

The European congeners of this plant, *Œ. fistulosa* and *Œ. crocata*, are undoubtedly poisonous. Our species is eagerly eaten by cattle, a fact that becomes very evident in a swamp near the Presidio, where it grows in company with *Cicuta Californica*, the latter not being touched by the animals when *Œnanthe* is eaten up to the roots. It would be desirable to investigate the properties of the species.

### 10. Osmorrhiza Raf.  SWEET CICELY.

Umbels regularly compound. Cremocarp terete, elongate, angulate, sulcate, hispid. Vittæ 0. Flowers white.—♃.

1. O. BRACHYPODA Torr. Involucral bracts linear and acuminate, equaling the flowers. Saucelito. Summer.

2. O. NUDA Torr. Involucre and involucels small and caducous.—Common. Summer.

### 11. Sium L.  WATER-PARSNIP.

Umbels regularly compound. Cremocarp laterally compressed, oblong. Mericarp dorsally convex, plain on the face, 5-ribbed. Ribs filiform. Intervals more than 1-vittate. Flowers white.—♃.

1. S. CICUTÆFOLIUM Gmelin.—Aquatic. Baden Station. Summer. Probably poisonous.

## 12. Cicuta L. Water-Hemlock.

Umbels regularly compound. Cremocarp laterally contracted. Ribs flattened, intervals 1-vittate. Mericarp terete. Aquatic. Flowers white.—♃.

1. C. Californica Gray. Leaves pinnate, the lower sometimes 2-pinnate at base; leaflets serrate, the veinlets running into the teeth. Presidio. Summer.

2. C. Bolanderi Watson. All leaves 2-pinnate; leaflets serrate, the veinlets running into the sinuses; involucral bracts linear.—Suisun. Alvarado marshes. Summer.

3. C. maculata L. Lower leaves 2-pinnate; leaflets serrate, the veinlets running into the sinuses; involucre obsolete.—Tamalpais. Summer. Poisonous aquatics.

## 13. Berula Koch.

Umbels regularly compound. Cremocarp laterally contracted, ovate. Mericarp terete, 5-ribbed, lateral ribs not contiguous. Epicarp thick, corky. Ribs filiform. Intervals ∞-vittate.

1. B. angustifolia Koch.—♃. Flowers white. Baden Station. Summer.

## 14. Pimpinella L.

Umbels regularly compound. Cremocarp laterally contracted, ovate. Mericarp convex, dorsally, plain on the ventral side, 5-ribbed. Ribs filiform. Lateral ribs contiguous. Intervals ∞-vittate.—♃.

1. P. APIODORA Gray. Flowers pale.—Contra Costa. Belmont. San Mateo. Summer.

Odor pleasant, like celery. An infusion of the root is used in cases of chronic catarrh, like the tincture of its European congener. *P. Saxifraga.*

### 15. Carum L. CARAWAY.

Umbels regularly compound. Cremocarp laterally compressed, oblong. Mericarp equally 5-ribbed. Ribs filiform. Lateral ribs contiguous. Commissure plain. Intervals 1-vittate. Flowers white.—⊙⊙.

1. C. KELLOGGII Gray.—Presidio. Alameda. Sunnyside. Summer. Seeds very fragrant.

### 16. Apiastrum Nutt.

Umbels regularly compound. Cremocarp very much contracted at the commissure, cordate. Mericarp incurved at base and apex, 5-ribbed; ribs little elevated. Intervals 1-vittate. Flowers white.—⊙.

1. A. ANGUSTIFOLIUM Nutt.—Niles. Spring.

### 17. Conium L. HEMLOCK.

Umbels regularly compound. Cremocarp laterally compressed. Endosperm deeply suliate on the ventral side. Mericarp 5-ribbed; ribs undulate, crenate. Vittæ 0. Flowers white.—⊙⊙.

1. C. MACULATUM L.—San Mateo. Sunnyside. Summer.

A well known medicinal and poisonous herb.

### 18. Deweya Torr. & Gray.

Umbels regularly compound. Cremocarp laterally compressed. Mericarp reniform in transverse section, 5-ribbed. Intervals more than 1-vittate. Flowers yellow.—♃.

1. D. HARTWEGI Gray. Ribs of mericarp prominent.—Oakland hills. Wild-Cat Creek. Strawberry Valley. Spring.

2. D. KELLOGGII Gray. Ribs of mericarp filiform.—Tamalpais. Spring.

### 19. Sanicula Tourn. SANICLE.

Umbels not regularly compound. Lobes of calyx limb foliaceous. Flowers polygamous. Cremocarp subglobose, aculeate. Ribs 0. Vittæ ∞.—♃.

1. S. TUBEROSA Torr. Leaves 2–3-pinnate; flowers yellow.—Tiburon. Spring.

2. S. BIPINNATIFIDA Dougl. Ripe cremocarp sessile; leaves long petiolate, triangular in outline, pinnately 3–5-lobed; segments distant, incisely lobed and decurrent on their rachis; flowers purple, in some localities yellow. Common. Spring.

3. S. LACINIATA Hook. & Arn. Ripe cremocarp sessile; leaves triangular in outline, 3-parted, the partitions pinnatifid to 2-pinnatifid. flowers yellow.—Tamalpais. Spring,

4. S. MENZIESII Hook. & Arn. Ripe cremocarp pedicillate; leaves deeply 3–5-lobed; the lobes broad, dentate; flowers yellow.—Common. Spring.

5. S. ARCTOPOIDES Hook. & Arn. Ripe cremocarp pedicillate; leaves deeply 3-lobed, the lobes laciniately cleft; flowers yellow.—Common. Spring.

6. S. MARITIMA Kellogg. Lower leaves entire, or slightly 3-lobed; upper leaves palmately parted, partitions cuneate and somewhat lobed; flowers greenish.—Alameda marshes. Summer.

### 22. Eryngium Tourn. ERYNGO.

Umbels not regularly compound. Lobes of the calyx-limb rigid. Cremocarp tuberculate. Ribs 0. Vittæ 0.—Spinose plants.

1. E. PETIOLATUM Hook. Flowers bluish. ⊙. Marshes. Summer.

### 21. Bowlesia Ruiz & Pavon.

Umbels simple. Cremocarp ovate, much contracted on the commissure, plane on the dorsal sides of the mericarps. Ribs 0. vittæ 0. Leaves opposite.

1. B. lobata Ruiz & Pavon. Flowers white.—☉. Golden Gate Park. San Mateo. Spring.

### 22. Hydrocotyle Tourn. Water Pennywort.

Umbels simple. Cremocarp laterally compressed, c a r i n a t e. Ribs filiform. Vittæ 0. Aquatic.—♃.

1. H. prolifera Kellogg. Leaves peltate, emarginate at base, crenate. Flowers greenish.—San Francisco. Summer.

2. H. ranunculoides L. Leaves orbicular, not peltate, lobate; lobes crenate; flowers greenish.—San Francisco. Summer.

### Family 2. ARALIACEÆ.

Fruit a berry or drupe. Leaves alternate.

#### 1. Panax L. (*Fatsia* Benth. & Hook).

Polygamous. Petals 5. Stamens 5. Ovary 2-celled. Styles 2. Fruit a 2-celled berry.

1. P. horridum Benth. & Hook. Flowers greenish.— ♄. Tamalpais. Summer.

#### 2. Aralia L. Spikenard.

Ovary 5-celled. Styles 5. Fruit a drupe, with five pyrenæ.

1. A. californica Watson. Flowers greenish; berry dark purple or black.—♃. Shady gulches. Summer.

## Family 3. CORNACEÆ.

Stamens 4. Leaves opposite.

### 1. Cornus L.  Cornel. Dogwood.

Flowers ☿, Petals 4. Ovary 2-celled. Style 1. Fruit a drupe, with two pyrenæ.— ♄.

1. C. Nuttallii Audubon. Flowers in a dense glomerule, supported by a conspicuous petaloid involucre; flowers greenish; fruit red.—Formerly in the vicinity of San Francisco, where it is now extinct. Still to be found on Bolinas Heights. Common in the Sierra.

2. C. Californica C. A. Meyer. Flowers in a cyme, not involucrate white; fruit white, sometimes blue.—Crystal Springs. Marin County. Summer.

### 2. Garrya Dougl.  Tassel Tree.

Diœcious. Flowers amentaceous, ternate between decussate bracts. Petals 0. Ovary 1-celled; styles two, persistent. Ovules two. Fruit a berry.— ♄.

1. G. elliptica Dougl. Leaves sessile, undulate; fruit purple.—Shady ravines around the bay. Spring.

2. G. Fremontii Torr. Leaves petiolate, not undulate; fruit purple.—Wright's Station. Santa Cruz mountains. Spring. "Quinine

tree" of the settlers. All parts intensely bitter; even the berries, which in *G. elliptica* are sweet and edible, although the bark and leaves partake of the bitterness of the other species.

ORDER 2. CORNICULATÆ. Calyx gamosepalous; stamens equaling or double the number of the petals: always correspondent. Ovaries compound of several carpidia, each carpidion ∞-ovulate.

### Family 1. SAXIFRAGACEÆ.

Number of carpidia less than that of the petals. Fruit a capsule.

#### 1. Saxifraga L. SAXIFRAGE.

Stamens 10. Pistils two. Ovary 2-celled. Fruit a loculicidal capsule.—♃.

1. S. VIRGINIANA Michx. Petals white.—Lagunitas Creek. Sunnyside. Spring.

#### 2. Boykinia Nutt.

Stamens five. Pistils two. Ovary 2-celled. Fruit a 2-celled capsule.—♃.

1. B. OCCIDENTALIS Torr. & Gray. Petals white.—Lagunitas Creek. Summer.

#### 3. Tellima R. Brown.

Stamens 10. Pistils 2-3. Petals lobed. Ovary 1-celled; styles short; stigmas capitate. Capsule valvularly dehiscent near the apex. —♃.

1. T. GRANDIFLORA R. Br. Petals laciniately pinnatifid, purple.—Saucelito. Tamalpais. San Miguel. Spring.

2. T. BOLANDERI Gray. Petals almost entire; styles smooth; calyx almost hypogynous; petals white. Tamalpais. Spring.

3. T. HETEROPHYLLA Hook. & Arn. Petals obtusely 3-lobed; styles smooth; calyx almost hypogynous; petals white.—Common. Spring.

4. T. AFFINIS Bolander. Petals 3-dentate; styles granulose; calyx perigynous; petals white.—San Rafael. Spring.

### 6. Tiarella L.

Stamens 10. Pistils 2. Petals entire. Ovary 1-celled; styles long; stigmas simple. Capsule valvularly dehiscent to the base. Valves unequal. Placentæ parietal. Flowers white. ♃.

1. T. UNIFOLIATA Hook.— Marin County. Spring.

### 5. Heuchera L.  ALUM-ROOT.

Stamens 5. Pistils 2. Petals entire. Ovary 1-celled; styles long. Capsule valvularly dehiscent. Valves equal.—♃.

1. H. MICRANTHA Dougl. Calyx pointed at the base, shorter than its pedicel.—Camp Taylor. Summer.

2. H. PILOSISSIMA Fisch. & Mey. Calyx rounded at the base; about the length of its pedicel.—Camp Taylor. Tamalpais. Crystal Springs. Summer.

## Family 2. RIBESIACEÆ.

Stamens 5. Pistils 2. Fruit a berry. Leaves alternate.

### 1. Ribes L.
GOOSEBERRY. CURRANT.

Ovary inferior, 1-celled, with 2 parietal, ∞-ovulate placentæ.— ♃.

1. R. MENZIESII Pursh. Thorny; vernation plicate; anthers sagittate; calyx tube campanulate, purple; petals white; berry prickly, yellowish.—Common. Spring. Fruit scarcely edible. "Prickly Gooseberry."

2. R. DIVARICATUM Dougl. Thorny; vernation plicate; anthers not sagittate; calyx tube campanulate, dull colored; petals white; berry not prickly, dark purple.—Common. Spring. Fruit edible.

3. R. SANGUINEUM Pursh. Not thorny; vernation plicate; calyx tube cylindrical, rose-red; berry glandular, dark.—Common. Spring. "Wild Currant." Fruit scarcely edible.

4. R. AUREUM Pursh. Not thorny; vernation convolute; flowers yellow.—Rare. Strawberry Valley. Wild-Cat Creek. Spring. "Golden Currant."

## Family 3. PHILADELPHEÆ.

Ovary inferior or half inferior. Number of carpidia disposed to correspond with calyx lobes. Fruit a capsule. Leaves opposite.

### 1. Whipplea Torr.

Stamens 5. Pistils 3-5. Ovary 3-5-celled. Cells 1-ovulate. Fruit a septicidal capsule. Flowers white.— ♄.

1. W. MODESTA Torr.—Redwoods. Summer.

## Family 4. CRASSULACEÆ.

Calyx, corolla and stamens alternating. If the stamens form two circles the inner circle is opposite to the petals. Ovaries opposite the petals. Fruit follicles with central dehiscence.

### 1. Cotyledon L.
(*Echeveria* DC.)

Calyx 5-parted. Petals 5, coherent by their claws. Stamens 10.— ♃. Fleshy herb.

1. C. CÆSPITOSA Haworth. Flowers yellow. Rocky places. Summer.

### 2. Sedum L. STONE CROP.

Petals 5, entirely free. Stamens 10.— ♃. Fleshy herbs.

1. S. SPATHULIFOLIUM Hook. Leaves glaucous, obtuse, narrowed toward the base; flowers yellow. Rocks near the Presidio. Summer.

2. S. stenopetalum Pursh. Leaves acute, lanceolate; flowers yellow.— Camp Taylor. Summer.

### 3. Tillæa L.

Stamens as many as petals.—⊙. Minute vernal aquatics.

1. T. minima Miers. Flowers clustered, white.—San Miguel.

2. T. angustifolia Nutt. Flowers solitary, white.—Mission Dolores.

Section 2. Centrospermæ. Central placentation from the base of the ovary.

Order 1. FICOIDALES. Anomalous, transitional types.

### Family 1. FICOIDEÆ.

Calyx superior. Petals ∞. Stamens ∞-seriate, inserted with the petals. Ovary 4-20-celled, ∞-ovulate. Fruit a capsule, ∞-seeded. Stipules 0.

**1. Mesembrianthemum** L. Fig-Marygold.

Characters of the family.

1. M. æquilaterale Haworth. Flowers red.—♃. Rocks near the seashore. Summer. Fruit edible.

Order 2. CARYOPHYLLALES. Flowers regular. Calyx inferior. Sepals as many as petals. Stamens as many or twice as many. Ovary 1-celled, placenta central.

## Family 1. SILENEÆ.

Calyx gamosepalous. Petals and stamens inserted into a carpophore. Stamens if equal to the petals alternating with them. Ovary ∞-ovulate. Styles several. Fruit a capsule. Leaves opposite. Stipules 0.

### 1. Silene L.
CAMPION. CATCHFLY.

Calyx 5-dentate without scales at the base. Petals unguiculate. Stamens 10. Styles 3. Capsule dehiscent by teeth; seeds reniform.

1. S. GALLICA L. Villous; leaves spathulate; flowers nearly sessile, racemose; petals pale, entire, scarcely exceeding the calyx.—⊙. Common. Summer.

2. S. ANTIRRHINA L. Glabrous; leaves lanceolate; flowers in a leafless, dichotomous panicle on long pediclels; petals obovate, equaling the calyx.—⊙. Livermore. Summer.

3. S. CALIFORNICA Durand. Flowers few on the ends of branches; petals deeply parted, with bifid segments; scarlet.—⊙. Oakland waterworks. Summer.

4. S. LACINIATA Cav. Flowers paniculate; petals deeply 4-cleft, with linear, acute lobes; bright scarlet.—♃. Contra Costa range. Summer.

5. S. VERECUNDA Watson. Petals bifid, rose-colored.—♃. Cemetery of San Francisco. Summer.

### Family 2. ALSINEÆ.

Sepals distinct to the base. Carpophore 0. Stamens, if equaling the petals, alternating with them. Ovary $\infty$-ovulate; styles several. Fruit a capsule. Leaves opposite; stipules 0.

#### 1. Cerastium L.
MOUSE-EAR CHICKWEED.

Sepals 5. Petals 5, emarginate. Stamens 10. Stigmas 5. Capsule cylindrical, dehiscent by 10 teeth. Flowers white.

1. C. ARVENSE L. Petals about twice the length of the sepals.—♃. Common. All the year round.

This weed has, in Europe, the reputation of being poisonous to sheep.

2. C. PILOSUM Ledebour. Petals but little exceeding the sepals.—♃. Punta de los Reyes. Summer.

#### 2. Stellaria L.   CHICKWEED.

Petals 2-lobed. Stamens 10. Stigmas 3-5. Capsule globose, dehiscent by 6-10 valves. Flowers white.

1. S. MEDIA L. Flowers on slender pedicels.—☉. Common. All the year round.

Ballast weed, introduced from Europe.

2. S. NITENS Nutt. Flowers on short pedicels.—⊙. Tamalpais. Spring.

### 3. Arenaria L.   SANDWORT.

Sepals unchanged in fruit. Petals entire. Stamens 5 or 10. Stigmas 3. Capsule ovoid, dehiscent by 3 valves, these valves sometimes 2-parted. Flowers white.

1. A. DOUGLASII Torr. & Gray. Valves of the capsule entire; leaves filiform; sepals acute, 3-nerved.—⊙. Tamalpais. Summer.
2. A. CALIFORNICA Brewer. Valves of the capsule entire; leaves lanceolate; sepals acute, 3-nerved.—⊙. Oakland Hills. Spring.
3. A. PALUSTRIS Watson. Valves of the capsule entire; leaves linear; sepals obtuse, without nerves. Formerly in a marsh near San Francisco, where at present it is extinct. Has not been found in any other locality.
4. A. MACROPHYLLA Hook. Valves of the capsule 2-cleft; sepals acuminate, 1-nerved. ♃.—Saucelito. Summer.

### 4. Sagina L.   PEARLWORT.

Sepals 4 or 5. Petals 4 or 5. Stamens twice 4 or twice 5. Stigmas 4 or 5, alternate with the sepals. Capsule dehiscent in 4 or 5 valves, opposite to the sepals.—⊙.

1. S. OCCIDENTALIS Watson. Petals white. San Francisco. Spring.

## Family 3. PARONYCHIEÆ.

Like Alsineæ, but the parts frequently defective or reduced in numbers. Leaves with scarious stipules.

### 1. Spergula L.   Corn-Spurrey.

Stamens 5 or 10. Ovary $\infty$-ovulate; styles 5, alternate with the sepals. Capsule 5-valved. Valves opposite to the sepals.—☉.

1. L. ARVENSIS L. Flowers white. — San Francisco. Summer. Ballast weed, introduced from Europe.

### 1. Spergularia Pers. (*Lepigonum* Fries).
Sand-Spurrey.

Ovary $\infty$-ovulate; style 3-5-cleft. Capsule 3-5-valved.

1. S. MACROTHECA Fisch. & Meyer. Leaves with large ovate stipules; flowers rose-colored.—♃. Margins of salt marshes. Summer.

2. S. MEDIA Fries. Leaves with short stipules; flowers white.—☉. Contra Costa. Spring.

### 3. Pentacæna Bartl.

Divisions of the calyx unequal, persistent in fruit; the three external with cucullate apex, ending in a spine; two internal mucronate. Petals 5, minute. Stamens 3-5. Ovary 1-ovulate; style 2-cleft. Fruit a utricle.—☉.

1. P. RAMOSISSIMA Hook. & Arn. Prostrate; pungent; everlasting.—♃.Presidio. Cemetery. Spring.

## Family 4. PORTULACACEÆ.

Flowers regular, but parts not corresponding in numbers.

### 1. Portulaca Tourn.   PURSLANE.

Calyx tube connate with ovary. Limb 2-parted, free and circumscissile. Petals 4-6. Stamens 8 to ∞. Capsule circumscissile.

1. P. OLERACEA L. Fleshy herb; petals yellow.--☉. San Francisco. Spring. Summer. Escaped from cultivation.

### 2. Calandrinia HBK.

Sepals 2, persistent. Petals 5 to ∞, equal. Stamens opposite to the petals, variable in number. Ovary ∞-ovulate; style 3-cleft. Capsule 3-valved, ∞-seeded.

1. C. MENZIESII Hook. Petals red.—☉. Common. Spring.

### 3. Claytonia L   MINER'S LETTUCE.

Sepals 2, persistent. Petals 5, equal. Stamens 5, opposite to the petals. Ovary 3 or 6-ovulate; style 3-cleft. Capsule 3-valved, 3-seeded.

1. C. SIBIRICA L. Inflorescence in loose, simple racemes. Herbaceous bracts with most

of the pedicels; single pair of cauline leaves distinct; leaves thin; petals rose-color.—♃. Saucelito. Spring.

2. C. PERFOLIATA Don. Inflorescence fasciculate; bracts few and minute; leaves thick, succulent; single pair of cauline leaves entirely connate; petals white.—☉. Common. Spring. The whole plant edible as a salad.

3. C. PARVIFLORA Dougl. Inflorescence becoming at last racemose; bracts few and minute; leaves thick, succulent; single pair of cauline leaves imperfectly connate, sometimes on one side entirely distinct; petals pale rose-color.—☉. Berkeley. Spring.

4. C. SPATHULATA Dougl. Inflorescence in a loose raceme; bracts few and minute; leaves thick, succulent; single pair of cauline leaves lanceolate but little connate, frequently entirely distinct; petals bright rose-color.—☉. Tiburon. Tamalpais. Spring.

5. C. CHAMISSONIS Eschsch. Several pairs of cauline leaves; petals white.—☉. Tiburon. Spring.

6. C. LINEARIS Dougl. Leaves alternate. Petals white.—☉. Guerneville. Spring.

### 4. Montia L.

Sepals 2, persistent. External petals 3; internal petals 2, connate. Stamens 3 or 5. Ovary 3-ovulate; style 3-cleft. Capsule 3-valved, 3-seeded. Flowers white.—☉.

1. M. FONTANA L. Flowers inconspicuous.—Sunnyside. Spring.

ORDER 3. CHENOPODIALES. Corolla 0, Perigonium inferior. Stamens opposite to the sepals. Ovary 1-celled, centrospermous. Stipules 0.

### Family 1. AMARANTACEÆ.

Perigonium without tube, 3-bracteate; anterior bract longer than the two lateral ones. Style simple.

#### 1. Amarantus L.

Polygamous, monœcious. Stamens distinct. Style short; stigmas 2 or 3. Fruit a circumscissile utricle.—⊙.

1. A. RETROFLEXUS L. Flowers green.—Cultivated grounds. Summer.

### Family 2. CHENOPODIACEÆ.

Perigonium without tube, not more than 1-bracteate.

#### 1. Chenopodium L.    PIG WEED.

Bracts 0. Perigonium 5-cleft. Lobes dorsally carinate. Stamens 5. Ovary 1-celled, 1-ovulate; stigmas 2. Fruit a depressed utricle enclosed in the persistent perigonium.

1. C. ALBUM L. Leaves smooth, pruinose, rhombic, sinuate; the uppermost lanceolate,

entire; flowers densely clustered in dense spikes, forming a strict, close panicle.—☉. Cultivated grounds. Summer.

Ballast weed, introduced from Europe.

2. C. MURALE L. Leaves shining, rhombic, dentate; flowers clustered in loose spicate panicles.—☉. Cultivated grounds. Summer.

Ballast weed, introduced from Europe.

3. C. AMBROSIOIDES L. Leaves lanceolate, dentate; flowers clustered in slender axillary spikes.—☉. Cultivated grounds. Summer.

Introduced from South America. Possesses anthelminthic properties.

4. C. CALIFORNICUM Watson. Leaves triangular, hastate; flowers densely clustered in simple terminal spikes.—♃. Cultivated grounds. Summer.

### 2. Atriplex L. ORACHE.

Flowers polygamous. ♂ and ☿ perigonium 3 or 5-parted. ♀ perigonium 2-parted. Stigmas 2. Fruit a compressed utricle.

1. A. PATULUM L. Leaves 3-angular, hastate; lower leaves opposite; bracts large, rhombic to hastate.—☉. Cultivated grounds. Summer. Ballastweed, introduced from Europe.

2. A. CORONATUM Watson. All leaves alternate; bracts orbicular, surrounded by a herbaceous, dentate margin.—☉. Contra Costa. Summer.

3. A. LEUCOPHYLLUM Dietr. All leaves alternate; fruiting bracts spongy, rhombic to ovate. ☉.—On the seashore. Summer.

4. A. CALIFORNICUM Mag. Lower leaves opposite, sessile; stem and leaves furfuraceous; fruiting bracts membranaceous; rhombic.—♃. On the seashore. Summer.

### 3. Salicornia Tourn. SAMPHIRE.

Flowers in threes, immersed in a rachis, decussately arranged and forming a spike. Perigonium gamophyllous, saccate. Stamens 1 or 2. Fleshy, articulate, aphyllous, saline plants.

1. S. AMBIGUA Michx.—♃. Salt marshes. Summer.

### 4. Suæda Forsk. SEA BLITE.

Flowers axillary, with minute, scale-like bracts. Perigonium urceolate, 5-cleft. Stamens 5. Fruit a utricle, enclosed in the inflexed perigonium. Leaves terete, fleshy.

1. S. CALIFORNICA Watson—♃. Salt marshes on an island near Alameda. Summer.

### Family 3. NYCTAGINEÆ.

Flowers involucrate. Perigonium gamophyllous, corolla-like, its persistent tube enclosing the akene.

### 1. Abronia Juss.

Involucre 5-leaved, ∞-flowered, persistent. Perigonium salver-shaped. Stamens 5, included. Stigma clavate. Leaves opposite.

1. A. UMBELLATA Lam. Leaves attenuate into a slender petiole; perigonium rose-colored.—♃. Seashore. Summer.

2. A. LATIFOLIA Esch. Petioles distinct, but short; perigonium yellow.—♃. Seashore. Summer.

Section 3. EUCYCLICÆ. Floral parts distinct in well defined circles.

ORDER 1. TRICOCCÆ. Flowers diclinous. Ovary superior. Ovules 1–2 collateral, pendulous from the summit of cells, which separate at last from a central axis.

### Family 1. EUPHORBIACEÆ.

Cells 1-ovulate.

#### 1. Eremocarpus Benth.

Monœcious. Flowers cymose. ♂: perigonium 5-parted; stamens 6 or 7, central, inflexed in æstivation. ♀: perigonium 0; ovary with 5 glands at the base, 1-celled, 1-ovulate; style simple; capsule 2-valved.—☉.

1. E. SETIGER Benth.—☉. Marin County. Niles. Summer.

The herb has the smell of strawberries, but is poisonous. The crushed leaves are used by the Indians to catch fish by stupefying them.

## 2. Hendecandra Esch.
### (*Croton* L.)

Diœcious. ♂: flowers in racemes; perigonium 5-cleft, with 5 glands opposite to the lobes; stamens more than 5. ♀ flowers: perigonium, 5-cleft; glands 0; ovary 3-celled; styles 3, each 4-parted; capsule 3-coccous, each coccus 2-valved.

1. H. PROCUMBENS Esch. (*Croton Californicus* Mull. Arg.)—♃. Common in sandy soil. Summer. Drastic poison.

## 3. Euphorbia L.   SPURGE.

Monœcious. Androgynous. Flowers umbellate. Involucre campanulate. ♂: flowers, perigonium 0; monandrous, stipitate; bracteate. ♀: flower solitary in the center of the umbel; perigonium dentate or 0; ovary 3-celled; styles 3, each 2-cleft.

1. E. OCELLATA Dur. & Hilg. Glands of the involucre petaloid; leaves all opposite, entire.—⊙. Sonoma. Millbrae. Summer.

2. E. SERPYLLIFOLIA Pers. Glands of the involucre with a narrow, white margin; leaves all opposite; base oblique; apex denticulate. ⊙.—Marin County. Summer.

3. **E. leptocera** Engelm. Glands of the involucre crescent-shaped, without colored margin; cauline leaves scattered, those of the branchlets opposite.—☉. Common. Summer.

4. **E. lathyris** L. Glands of the involucre crescent-shaped, without colored margin; leaves all opposite.—☉☉. Common. Summer.

Poisonous plant introduced from Europe and found everywhere. The seeds contain an oil analogous to that of croton. This plant is said to drive away or kill rats, as also the castor-oil plant. *Ricinus communis*.

Order 2. **MALVALES.** Flowers regular. Calyx free, 5-parted, valvate in æstivation. Corolla 5-parted, contorted in æstivation. Stamens often monadelphous. Carpidia several, free, or connate with the central axis into an ∞-celled ovary. Leaves alternate, stipulate.

### Family 1. MALVACEÆ.

Claws of petals united with each other, and with the column of ∞, monadelphous stamens. Anthers reniform, 1-celled.

1. **Lavatera** L.  Tree Mallow.

Involucre 3-6-cleft. Ovaries ∞, verticillate; 1-ovulate. Style 1, springing from the central axis; stigmas ∞, filiform. Fruit a schizocarp; seeds ascending.

1. L. ASSURGENTIFLORA Kellogg. Petals rose-colored. — ♄. Near cultivated places. Summer.

Native of the island of Anacapa, but having escaped cultivation, now spontaneous in different localities.

## 2. Malva L. MALLOW.

Involucre 3-leaved. Ovary $\infty$-celled; cells 1-ovulate. Styles as many as cells, united at base; stigmas obtuse. Fruit a depressed capsule; seed ascending.

1. M. BOREALIS Wallman. Flowers pale.— ☉. Cultivated grounds. Summer. Ballast weed from Europe.

## 3. Sidalcea Gray.

Involucre 0. Ovary $\infty$-celled; cells 1-ovulate. Styles as many as cells, united at base. Fruit 5-10-coccous schizocarp; seed ascending.

1. S. HUMILIS Gray (*malvæflora?* Gray). Raceme long and loose; petals rose or purple. ♃. Common. Summer.

2. S. DIPLOSCYPHA Gray. Flowers in umbellate clusters; pedicels subtended by persistent, hispid bractlets; petals rose or purple. ☉. Suñol. Summer.

3. S. MALACHROIDES Gray. Flowers in close, racemose, nearly spicate clusters; petals small, white.—☉. Marin County. Summer.

### 4. Malvastrum Gray (*Sphæralcea* St. Hilaire).

Ovary ∞-celled; cells 2-ovulate. Styles as many as cells, united at base; stigmas capitate. Fruit a loculicidal capsule, at length also septicidal.

1. M. Thurberi Gray. Petals pale.— ♄. San Mateo. Niles. Summer.

### 5. Sida Kunth.

Ovary ∞-celled; cells 1-ovulate. Styles as many as cells, united at base; stigmas capitate. Fruit a ∞-seeded capsule. Seeds pendulous from the apex.— ♃.

1. S. hederacea Torr. Flowers pale. Common. Summer.

Order 3. GERANIALES. Calyx free, imbricated in æstivation. Petals 5, contorted or convolute in æstivation; stamens hypogynous; definite. Ovary the compound of a definite number of carpidia.

### Family 1. GERANIACEÆ.

Sepals 5. Petals 5; hypogynous; alternate with the sepals; stamens 10. Carpidia 5, verticillate round a columnar axis (gynophore). Styles distinct at base; connate towards their apex. Carpidia 5, 2-ovulate, 1-seeded, ventrally dehiscent. Leaves stipulate.

### 1. Geranium L.

Stamens monadelphous, all fertile. Styles persistent.

1. G. Carolinianum L. Corolla rose-color.—⊙. Common. Summer.

### 2. Erodium L'Her.
#### Storksbill. Pin-clover.

Stamens monadelphous, alternately sterile or depauperate. Styles persistent.

1. E. cicutarium L'Her. Leaves pinnate; leaflets pinnatifid, with narrow acute lobes; flowers rose-colored.—⊙. Common. Summer. Fodder plant, called Alfilerilla (*alfiler*, in Spanish, needle) from the shape of the fruit. If the plant is not indigenous, it must have been introduced for a considerable time, or else it would not have a distinct name from the old settlers.

2. E. moschatum L'Her. Leaves pinnate; leaflets ovate, doubly serrate; stipules conspicuous; flowers rose-color.—⊙. Cultivated grounds. Summer. Ballast weed, introduced from Europe.

3. E. botrys Bertol. Leaves pinnatifid, lobes dentate, stipules small; petals pale.—⊙. Cultivated grounds. Summer.

4. E. macrophyllum Hook. & Arn. Leaves reniform, lobed.—⊙. Niles. Summer.

## Family 2. OXALIDEÆ.

Sepals 5. Petals hypogynous. Ovary 5 carpidia, opposite the petals, attached to the axis by their central angle, 2-to-∞-ovuled; ovules vertically arranged. styles 5, persistent. Leaves alternate.

### 1. Oxalis L.  Wood-Sorrel.

Petals 5. Stamens 10, monadelphous, the 5 opposite the petals shorter. Capsule oblong.

1. O. Oregana Nutt. Peduncles 1-flowered; petals white or rose-color.—♃. Crystal Springs. Camp Taylor. Spring.

2. O. corniculata L. Peduncles two to several-flowered; petals yellow.—♃. Common. Spring.

## Family 3. LINEÆ.

Calyx persistent. Petals hypogynous, unguiculate. Perfect stamens 5, sometimes 4. Ovary 5, sometimes 4, 3 or 2-celled; cells 2-ovulate. Ovules collateral, more or less separated by a spurious septum. Styles equalling the cells in number. Capsule globose.

### 1. Linum.  Flax.

Sepals 5. Petals 5.

1. L. perenne L. Pentagynous; petals blue.—♃. Colma. Millbrae. Summer.

2. L. Breweri Gray. Trigynous; pedicels short; stipular glands conspicuous; flowers almost solitary; petals yellow.—⊙. Contra Costa hills. Summer.

3. L. congestum Gray. Trigynous; pedicels short; stipular glands very small; sepals pubescent; flowers in terminal fascicles; petals dark-red.—⊙. Tamalpais. Summer. Rare.

4. L. Californicum Benth. Trigynous; pedicels short; stipular glands conspicuous; upper flowers cymose, lower flowers solitary; petals rose-colored.—⊙. San Francisco cemetery. San Rafael. Summer.

5. L. spergulinum Gray. Trigynous; pedicels conspicuous; stipular glands 0; flowers rose-colored or white.—⊙. Marin County. Summer.

### Family 4. LIMNANTHEÆ.

Calyx persistent, valvate in æstivation. Petals alternating with calyx lobes, inserted into a perigynous disc. Stamens twice as many as petals. Carpidia verticillate, free, 1-ovulate; style central on the apex, 5, sometimes 3-cleft. Fruit 5, sometimes 3 akenes. Stipules 0.

#### 1. Limnanthes R. Br.

Calyx 5-parted. Petals 5, cuneiform. Ovaries 5. Fruit 5, rugose akenes.—⊙.

1. L. Douglasii R. Br. Sepals glabrous; petals white or rose-colored, claws always yellow. Sunnyside. Summer.

2. L. alba Hartweg. Sepals villous; petals white, claws yellow.—Millbrae. Summer.

Order 4. TEREBINTHALES. Flowers regular. Calyx free. Petals 5, imbricate or valvate in æstivation, not contorted, nor truly convolute, inserted into a disc. Stamens 5, or a multiple. Ovary 1–5 carpidia, syncarpous or apocarpous; carpidia 1–2-ovulate.

### Family 1. RUTACEÆ.

Stamens inserted on the external margin of the disc. Carpidia more than 1-ovulate. Stipules 0.

#### 1. Ptelea L.  Hop-tree.

Polygamous. Calyx 4–5-parted. Petals 4 or 5. Stamens 4 or 5, alternate with the petals. Ovary on a convex disc, 2-celled; cells 2-ovulate; style 1. Fruit a 2-seeded samara. Flowers white or greenish.— ♄ .

1. P. angustifolia Benth.—Niles. Spring.

### Family 2. TEREBINTHACEÆ. (*Anacardiaceœ*).

Stamens inserted on the inner margin of the disc. Ovary 1, or if more, only one fertile; 1-ovulate. Fruit indehiscent.

### 1. Rhus L.
POISON-OAK. YEDRA.

Polygamous. Calyx 5-parted, persistent. Stamens 5. Ovary one, 1-celled, 1-ovulate; styles 3. Fruit a dry drupe.

1. R. DIVERSILOBA Torr. & Gray. Flowers paniculate; petals greenish. Fruit pale.— ♄. Common. Summer.

2. P. AROMATICA Ait. Flowers spicate; petals yellow; fruit red.— ♄. Livermore. Spring.

Juice and exhalation of this shrub unlike those of R. diversiloba do not produce eczema, but are entirely harmless.

### Family 3. JUGLANDEÆ.
(Connecting link between *Terebinthales* and *Cupuliferæ*.)

Flowers diclinous. ♂: amentaceous; perigonium adnate to the bract, and imbricate in æstivation; petals 3, or its multiple. ♀: aggregate or racemose; perianth connate with the imperfectly 2–4-celled, 1-ovulate ovary; fruit a 1-pyrenous drupe; pyrena of valvular dehiscence. Leaves imparipinnate, stipules 0.

### Juglans L.    WALNUT.

♂: stamens more than six. ♀ flowers: few, terminal; calyx 4-parted; petals 4; style 2.— ♄.

1. J. CALIFORNICA Watson. — Very local. Walnut Creek. Spring.

ORDER 5. SAPINDALES. Parts of the androecium not symmetrical. Calyx free, imbricate in æstivation. Petals inserted into a hypogynous disc. Stamens generally more than petals, but not their multiple. Carpidia 3, sometimes 2, more or less connate into an ovary. Ovules of definite number.

### Family 1. POLYGALACEÆ.

Flowers irregular, often imitating *Papilonaceæ*. Sepals 5, the two lateral petaloid (wings). Petals 5, sometimes 3, connate with the stamineal tube. Anterior petal concave (carina). Stamens 8, sometimes 4 or less, usually monadelphous. Ovary 2-celled; style 1. Stipules 0.

#### 1. Polygala L.

Calyx persistent. Stamens 8, ascending. Filaments united at base into an anteriorly cleft tube. Ovary 2-celled; cells 1-ovulate. Capsule loculicidal.

1. P. CUCULLATA Benth. Flowers rose-color.—♃. Tamalpais. Crystal Springs. Summer.

### Family 2. SAPINDACEÆ.

Sepals 5, often irregular and more or less connate. Disc fleshy. Petals alternate with sepals, and appendiculate at their claws— sometimes one or all of them wanting. Stamens 1-seriate, more than 5. Ovary generally 3-celled; cells 1-2 ovulate.

### 1. Æsculus L. BUCKEYE.

Polygamous. Calyx irregular, gamosepalous. Petals unguiculate, 5, or by abortion of the anterior one, 4. Stamens 6-8, filaments ascending. Cells of ovary 2-ovulate. Fruit a capsule, generally by abortion 1-celled, 1-seeded. Leaves palmate, deciduous.— ♄.

1. Æ. CALIFORNICA Nutt. Petals white, tinged with red and yellow.—Common. Summer.

The seeds contain an active principle not yet sufficiently examined.

### 2. Acer Mœnch. MAPLE.

Polygamous. Petals as many as sepals, inserted into the margin of the lobate, hypogynous disc. Stamens 8, inserted with the petals. Ovary 2-celled, 2-lobed; cells 2-ovulate. Fruit a 2-seeded samara.— ♄. Leaves opposite, palmately lobed, deciduous.

1. A. MACROPHYLLUM Pursh. Flowers yellow.—San Mateo. Marin County. Spring.

### 3. Negundo Mœnch. BOX ELDER.

Polygamous. Diœcious. Petals 0. ♂: stamens 4, sometimes 5, opposite to the calyx teeth. ♀: ovary 2-celled, 2-lobed; cells 2-ovulate; fruit a 2-seeded samara. Leaves opposite, imparipinnate, deciduous.— ♄.

1. N. CALIFORNICUM Torr. & Gray.—San Mateo. Marin County. Spring.

ORDER 6. CELASTRALES. Parts of the flower correspondent in number. Disc tumid, adnate to the base of the calyx. Stamens inserted with the petals on the margin of the disc. Ovules definite.

### Family 1. RHAMNACEÆ.

Calyx conspicuous, gamosepalous, valvate in æstivation. Stamens inserted with the petals and alternate with the calyx lobes. Ovary 2-3-celled; cells 1-ovulate. Styles 2-3, more or less connate.

#### 1. Rhamnus Juss.

Calyx-tube urceolate, margin 4-5-cleft. Petals minute, with short claws, or 0. Ovary free. Fruit a 2-3-pyrenous drupe. Leaves alternate, stipulate.— ♄.

1. R. CROCEA Nutt. Flowers 4-merous, apetalous; fruit red.—Lake Chabot. San Rafael. Taylorville. Spring.

2. R. CALIFORNICA Esch. Flowers 5-merous; petals small, ovate, emarginate; ripe fruit black.—Common. Summer.

"*Cascara sagrada*" of the old Californians. An infusion of the bark of this species and of *R.*

*Purshiana* of Northern California was in use among them as a purge. There also was a belief that the infusion of the bark, if the branch was scraped towards the top, would act as an emetic.

An *infuso-decoct.* of the leaves is one of the most reliable remedies against fresh cases of the eczema caused by poison oak.

## 2. Ceanothus L.
### CALIFORNIA LILAC. CHAPPARAL.

Calyx-tube hemispherical, concave; margin petaloid. Petals unguiculate; margin cucullate. Ovary half immersed into the disc; style 3-cleft. Fruit a 3-seeded capsule with base embraced by the calyx tube, dehiscent at the inner angle of the cells. Leaves simple.— ♄.

1. C. THYRSIFLORUS Eschsch. Leaves all alternate, 3-nerved from their base; flowers in dense, compound racemes, blue, fragrant.— Common. Summer.

2. C. DENTATUS Torr. & Gray. Leaves all alternate, frequently fascicled, 3-nerved from the base; flowers in small fascicles on naked, terminal peduncles, blue.—Tamalpais. Summer.

3. C. DIVARICATUS Nutt. Leaves all alternate, 3-nerved from their base; flowers in racemes, pale; branches spinose.—Niles. Livermore. Summer.

4. C. papillosus Torr. & Gray. Leaves all alternate, pinnately veined, glandular; flowers blue.—Redwood City. Alma. Summer.

5. C. crassifolius Torr. Leaves often opposite, with numerous straight veins, oblong; young branches tomentose; flowers in dense fascicles, pale.—Wright's Station. Summer.

6. C. cuneatus Nutt. Leaves often opposite, cuneate, with numerous straight veins; young branches almost smooth; flowers in loose fascicles, pale.—Marin County. Summer.

### Family 2. CELASTRACEÆ.

Calyx conspicuous, gamosepalous. Stamens alternate with the petals and opposite to the calyx lobes.

#### 1. Euonymus Tourn.    Spindle-Tree.

Sepals and petals 4 or 5, spreading. Ovary 3-4-celled, half immersed into the disc. Capsule loculicidal.

1. E. occidentalis Nutt. Flowers dark brown.— ♄. Crystal Springs. Taylorville. Summer.

### Family 3. AMPELIDEÆ.

Calyx inconspicuous. Stamens inserted with the petals, opposite and of equal number.

Ovary 3-celled; cells 2-ovulate; ovules collateral. Fruit a berry. Lower leaves opposite; upper by the transformation of one of the pair into a tendril, apparently alternate.

### 1. Vitis Tourn.    Grape.

Petals coherent at their apex, caducous.

1. V. CALIFORNICA Benth.— ♄.

Order 7. GUTTIFERALES. Flowers regular. Calyx free, sepals 4 or 5, imbricate or 2-seriate, never regularly valvate in æstivation. Petals convolute in æstivation, sometimes 0. Stamens ∞, more or less connate at base, sometimes reduced in number. Carpidia several, connate into an ovary.

### Family 1. HYPERICINEÆ.

Sepals 2-seriate. Petals as many as sepals and alternate. Carpidia 3 or 5. Ovules ∞, 2-seriate in each. Styles 3 or 5, more or less united. Fruit a septicidal capsule. Leaves opposite. Stipules 0.

### 1. Hypericum L.    St. John's-wort.

Stamens polyadelphous; stigmas capitate. Flowers yellow.—♃.

1. H. CONCINNUM Benth. Leaves not clasping; stamens decidedly polyadelphous in three phalanges; petals longer than the sepals.— Tamalpais. Camp Taylor. Crystal Springs. Summer.

2. H. anagalloides Cham. & Schlecht. Leaves clasping; stamens rather polyandrous; petals somewhat shorter than the sepals.— Presidio. Saucelito. Lagunitas. Summer.

### Family 2. SALICINEÆ

Diœcious. Amentaceous. ♂ : perigonium 0; stamens central. ♀ : perigonium 0; ovary 2 carpidia, connate into one; ovules ∞; styles 2; capsule 1-celled, 2-valved, ∞-seeded; seeds with hairy arillus. Leaves deciduous, alternate, stipulate.

#### 1. Salix Tourn.    Willow.

Bracts entire.

1. S. lævigata Bebb. Stamens more than 2; petioles downy; aments coeval with leaves. ♄. Alms House. Spring.

2. S. lasiandra Benth. Stamens more than 2; petioles glandular at their upper ends; aments coeval with leaves. — ♄. Alameda. Spring.

3. S. longifolia Muhl. Diandrous; scales deciduous; aments coeval with leaves. — ♄. Common. Spring.

4. S. lasiolepis Benth. Diandrous; scales persistent, darker at the apex; capsules acute, smooth, dark green; aments precocious.— ♄. Common. Spring.

5. S. COULTERI Anders. Monandrous; aments in the axils of persistent leaves.— ♄. Marin County. Spring.

### 2. Populus Tourn.
POPLAR. COTTONWOOD.

Bracts fimbriate, laciniate.— ♄.

1. P. TRICHOCARPA Torr. & Gray.— ♄. Marin County. Niles. Spring.

ORDER 8. PARIETALES. Ovary 1-celled; placentation parietal; ovules $\infty$.

### Family 1. CISTINEÆ.

Calyx 2-seriate; 2 external sepals smaller than the 3 internal, contorted to the left in æstivation, sometimes 0. Petals 5, caducous, contorted to the right in æstivation. Stamens $\infty$. Ovary free, formed by several connate carpidia; style 1. Fruit a capsule; placentation sometimes apparently axillary, by placentæ protruding towards the center.

### 1. Helianthemum Tourn. ROCK-ROSE.

Ovary 3-valved. Placentæ in the middle of the valves; each valve more than 2-seeded. Flowers yellow.

1. H. SCOPARIUM Nutt.— ♃. Tamalpais. Summer.

### Family 2. LOASACEÆ.

Calyx-tube more or less connate with the

ovary, costate; margin 4 or 5-parted. Petals 4 or 5, and their multiples. Stamens ∞, some of the external sometimes petaloid. Ovary inferior, 1-celled; ovules ∞; style 1. Fruit a capsule immersed in the calyx tube.

### 1. Mentzelia L.

Calyx limb 5-parted, persistent. Petals not cucullate. Stamens ∞, inserted in the throat of the calyx below the petals. Style 3-cleft. Capsule dehiscent near the apex.

1. M. LINDLEYI Torr. & Gray. Petals obovate, abruptly acuminate, yellow.—⊙. Niles. Suñol. San Leandro. Summer.

2. M. LÆVICAULIS Torr. & Gray. Petals lanceolate, cream-colored.—⊙ ⊙. Arroyo del Monte, near Livermore. Summer.

### Family 3. DATISCEÆ.

Calyx tube connate with ovary. Petals 0. Ovary 1-celled; placentæ on the middle nerve of the carpidia; ovules ∞; styles on the margins of the carpidia, 2 for each placenta, sometimes united into a contiguous style. Fruit a capsule, crowned by the calyx margin and style.

### 1. Tricerastes Presl. (*Datisca* L).

Flowers ☿. Calyx-tube 3-angular; margin 3-dentate. Stamens 3, alternate with the lobes of the calyx; anthers extrorse. Ovary with

3 placentæ; styles 3, 2-parted, opposite the calyx-teeth. —♃.

1. T. GLOMERATA Benth & Hook.—Sunnyside. Niles. Summer.

### Family 4. FRANKENIACEÆ.

Flowers regular. Calyx gamophyllus free, tubular, costate, persistent. Petals alternate with the lobes of calyx, unguiculate. Stamens not correspondent in number to petals; anthers extrorse. Ovary 1-celled. Fruit a capsule 2-4-valved, included in the persistent calyx-tube.

#### 1. Frankenia L.

Leaves opposite. Stipules 0.—♃.

1. F. GRANDIFLORA Cham. & Schlecht. Petals pink.—Seashore. Summer.

### Family 5. VIOLACEÆ.

Sepals 5, persistent. Petals 5, unguiculate. Stamens inserted into a hypogynous disc, filaments flat; anthers short, introrse, pressed against the ovary, the connective elongated beyond the anthers. Ovary free, 1-celled, formed by 3 carpidia; ovules ∞, 2-seriate; style 1, persistent; fruit a 3-valved capsule; valves bearing the seeds on the middle nerve. Leaves stipulate.

### 1. Viola L.  Violet.

Sepals unequal, appendiculate at base. Petals unequal, the inferior longest and calcarate. Stamens 5; anthers cohering into a tube, the two inferior ones appendiculate at their base.

1. V. ODORATA L. Stemless; petals violet. ♃. Throckmorton Ranch. Spring.

Sweet violet; probably escaped from cultivation.

2. V. CANINA L. Stem leafy; spur as long as the sepals; petals violet.—♃. Cemetery, Saucelito. Spring.

3. V. OCELLATA Torr. & Gray. Stem erect, leafy. Leaves cordate; stipules small, scarious; spur merely saccate; petals gradually changing color from purple to white.—♃. Tamalpais. Santa Cruz Mountains. Summer.

4. V. PEDUNCULATA Torr. & Gray. Stem leafy, ascending; leaves rhombic; stipules narrowly lanceolate, foliaceous; sepals lanceolate; spur merely saccate; petals yellow, the outside tinged with purple. — ♃. Common. Spring.

5. V. AUREA Kellogg. Stem leafy, ascending; leaves lanceolate; stipules lanceolate, foliaceous; sepals linear, acuminate; spur merely saccate; petals yellow.—♃. Contra Costa mountains. Summer.

6. V. SARMENTOSA Dougl. Stem prostrate, leafy; leaves cordate, finely crenate; petals yellow.—♃. Tamalpais. Taylorville. Summer.

7. V. LOBATA Benth. Erect leafy stem; cauline leaves palmate; petals yellow, externally tinged with purple.—♃. Marin County. Summer.

Series 2. APHANOCYCLICÆ. Floral parts showing a tendency to spiral arrangement, and generally distinct. Parts of the gynœcium only occasionally connate. Calyx and corolla often of the same or similar structure. Numerical law not yet established. Stamens generally more numerous than sepals and petals.

ORDER 1. CRUCIFLORÆ. Calyx and corolla formed from 2 and its multiples. Circles of stamens more than 1. Carpidia 2 and its multiples.

### Family 1. CRUCIFERÆ.

Sepals 2 pairs. Petals 4. Stamens tetradynamous. Ovary free, compound of 2 lateral carpidia; placentæ parietal, on the margins of a spurious septum (replum); style 1. Fruit a siliqua, silicula or lomentum. Leaves alternate; stipules 0.

**1. Raphanus** Tourn.   RADISH.

Indehiscent siliqua irregularly moniliform, several-seeded.

1. R. sativus L.—☉. Common all the year round.

Escaped from gardens. Originally a native of China, now in cultivation over the whole earth.

### 2. Thysanocarpus Hook.

Indehiscent silicula (samara) orbicular, winged, 1-seeded.—☉. Spring.

1. T. pusillus Hook. Pubescent throughout; cauline leaves lanceolate, sessile, but not clasping; silicula orbicular, hirsute.—Niles.

2. T. laciniatus Nutt. Cauline leaves linear, scarcely auricled at their clasping base; silicula obovate, slightly pubescent.—Niles.

3. T. curvipes Hook. Glabrous; cauline leaves conspicuously auricled at the clasping base; silicula orbicular.—Common.

### 3. Senebiera Poir.

Indehiscent silicula (schizocarp) 2-seeded. Seeds globose, rugose. Flowers minute, white.

1. S. didyma Pers.—☉. Common. Summer.

### 4. Lepidium R. Br. Peppergrass.

Silicula dorsally compressed, ovate, with carinate valves and narrow replum. Seeds solitary in each cell, pendulous from the apex of the replum.

1. L. NITIDUM Nutt. Silicula shining, glabrous; petals whitish.—⊙. Common. Spring.

2. L. LATIPES Hook. Silicula strongly reticulated; petals greenish.—⊙. Marin County. Summer.

### 5. Capsella Vent. SHEPHERD'S-PURSE.

Silicula dorsally compressed, with carinate, 1-nerved valves and narrow replum. Seeds several. Flowers white.—⊙.

1. C. BURSA-PASTORIS L.—Common all the year round.

Ballast weed, introduced from Europe.

### 6. Tropidocarpum Hook.

Siliqua dorsally compressed, linear, with 1-nerved, carinate valves, and very narrow replum. Seeds ∞. Flowers yellow.—⊙

1. T. GRACILE Hook.—Marin county. Summer.

### 7. Nasturtium R. Br. WATER-CRESS.

Siliqua linear, with concave valves, and only rudimentary nerve. Seeds ∞, 2-seriate in each valve.—Aquatic.

1. N. OFFICINALE R. Br. Petals white.—♃. Common all the year round.

Introduced from Europe, at present perfectly naturalized.

### 8. Sisymbrium L.  HEDGE-MUSTARD.

Siliqua linear, with concave, 3-nerved valves. Seeds ∞, 1-seriate in each valve.

1. S. OFFICINALE Scop. Branches divaricate; siliqua closely appressed to the axis of the raceme; petals yellow.—⊙. Common all the year round.

Ballast weed from Europe.

2. S. ACUTANGULUM DC. Branches ascending; siliqua erect; petals yellow.—⊙. Mission Dolores. San Jose. Summer.

Ballast weed, native of the Mediterranean region.

3. S. REFLEXUM Nutt. Branches but few; siliqua deflexed; petals yellow, white or red.—⊙. Presidio. Saucelito. Alameda. Summer.

### 9. Cakile Tourn.  SEA-ROCKET.

Fruit a 2-articulate lomentum.

1. C. AMERICANA Nutt. Petals rose-colored. ⊙. Berkeley salt marshes. Summer.

### 10. Barbarea R. Br.  WINTER-CRESS.

Siliqua linear, 4-angular with concave, 1-nerved valves. Seeds ∞, compressed, 1-seriate in each valve. Flowers yellow.

1. B. VULGARIS R. Br.—♃. San Mateo. Summer.

## 11. Brassica L.  MUSTARD.

Siliqua linear, with concave, more than 1-nerved valves. Seeds ∞, globose, 1-seriate in each valve. Flowers yellow.

1. B. CAMPESTRIS L. Upper leaves clasping, auriculate.—☉. Common. Summer.

Native of Europe, but escaped from cultivation and naturalized.

2. B. NIGRA Boiss. All leaves petioled.—☉. Very common. Summer.

Black mustard. Probably like the former escaped from cultivation, but at present a troublesome weed, plentiful enough to impart a disagreeable taste to milk and honey.

## 12. Erysimum L.

Siliqua linear, 4-angular, with carinate, 1-nerved valves. Seeds ∞, globose, 1-seriate in each valve. Flowers yellow.

1. E. ASPERUM DC. ☉☉. Common. Summer.

## 13. Caulanthus Watson.

Sepals equally saccate. Petals undulate. Siliqua elongated, terete, sessile; valves 1-nerved. Seeds ∞, oblong, 1-seriate in each valve. ☉☉.

1. C. PROCERUS Watson. Petals greenish white. Common. Summer.

### 14. Cheiranthus L.  WALL-FLOWER.

Lateral sepals saccate. Petals unguiculate. Siliqua laterally compressed; replum broad; valves 1-nerved. Seeds compressed, 1-seriate in each valve.

1. C. ASPER Cham. & Schlecht. Petals in different localities of different coloration, viz., near San Francisco cream-color, at San Mateo orange-color, in Saucelito light yellow.—☉☉. Common. Spring.

### 15. Streptanthus Nutt.

Sepals colored and equal. Petals undulate. Siliqua laterally compressed; valves flattened, 1-nerved. Seeds ∞, 1-seriate, marginate, compressed.

1. S. GLANDULOSUS Hook. Hispid; lateral pair of sepals not turned inward behind the upper petals; sepals and petals white or red.—☉. Sunnyside. Spring.

2. S. PERAMŒNUS Greene. Hispid; lateral pair of sepals turned inward behind the upper petals; sepals dark purple, petals pale.—☉. Wright's Station. Oakland Hills. Summer.

3. S. NIGER Greene. Glabrous, glaucous; sepals obtuse, dark purple; petals pale with purple claws.—☉. Tiburon. Spring.

### 16. Arabis L.  ROCK-CRESS.

Petals not undulate. Siliqua laterally compressed, linear; valves flattened, with prominent nerve. Seeds ∞, 1-seriate, compressed, marginate.

1. A. BLEPHAROPHYLLA Hook & Arn. Leaves conspicuously ciliate; cauline leaves oblong, sessile; petals purple.—♃. Presidio. Saucelito. Spring.

2. A. PERFOLIATA Lam. (*Turritis glabra* L.) Glaucous; cauline leaves ovate, clasping by the sagittate base; petals pale.—☉ ☉. Common. Summer.

### 17. Cardamine L.  BITTER-CRESS.

Siliqua laterally compressed, linear; valves flat, without nerve. Seeds ∞, 1-seriate; funicle filiform.

1. C. PAUCISECTA Benth. Lower leaves simple or 3-lobed, cauline leaves 5-foliolate.—♃. Common. Spring.

2. C. OLIGOSPERMA Nutt. All leaves pinnate. ☉. Saucelito. San Mateo. Spring.

### 18. Alyssum L.

Silicula laterally compressed, orbicular; valves concave; replum broad. Seeds 1 or 2 in each valve.

1. A. CALYCINUM L. Petals white or yellow, but little exceeding the sepals.—☉. San Francisco. Summer.

Native of the Mediterranean region, escaped from gardens,

2. A. maritimum L. Petals white, twice longer than the sepals.—♃. Oakland. Summer.

Native of the Mediterranean region, escaped from gardens.

### Family 2. FUMARIACEÆ.

Sepals 1 pair. Corolla irregular. Stamens 6 more or less connate, diadelphous.

#### 1. **Dicentra** Bork.    Bleeding-heart.

Sepals small, two lateral petals flat, anterior and posterior petals calcarate. Ovary 1-celled, with two parietal placentæ; $\infty$-ovulate. Fruit a 2-valved capsule.—♃.

1. D. chrysantha Hook. & Arn. Flowers yellow, panicled on a leafy stem.—Crystal Springs. Summer.

2. D. formosa DC. Flowers rose-color, racemose on a scape.—Colma. Taylorville. Summer.

### Family 3. PAPAVERACEÆ.

Sepals 2 or 3, caducous. Corolla regular; petals two pairs, sometimes three. Stamens not connate. Ovary free.

#### 1. **Platystemon** Benth.    Cream Cups.

Sepals 3. Petals 6. Stamens $\infty$; filaments flattened their whole length. Carpidia several, $\infty$-seeded, at first syncarpous; stigmas free, sessile, linear. Fruit at last apocarpous, dissolving into several follicles. Plant villous. Flowers cream color.—☉.

1. P. CALIFORNICUS Benth. Common. Spring.

### 2. **Platystigma** Benth. CREAM CUPS.

Sepals 3. Petals 6. Stamens ∞; filaments flattened only at the base. Ovary 3-angular, 1-celled; placentæ 3; stigmas 3. Fruit a capsule, 3-angular, 3-valved, ∞-seeded.—☉. External petals yellow, internal cream color.

1. P. LINEARE Benth. Filaments flat.— Common. Spring.
2. P. CALIFORNICUM Benth. & Hook. Filaments terete.—San Mateo. Spring.

### 3. **Dendromecon** Benth.

Sepals 2. Petals 4. Stamens ∞. Ovary terete, linear with 2 placentæ; ovules ∞; stigma sessile, 2-lobed. Fruit a capsule 1-celled, 2-valved; seeds on the margin of the valves. Flowers yellow. Leaves rigid, entire.— ♄.

1. D. RIGIDUS Benth.—Tamalpais. Wright's Station. Summer.

### 4. **Eschscholtzia** Cham. CALIFORNIA POPPY.

Calyx and corolla inserted into a turbinate thalamus. Calyx gamosepalus, circumscissile at base. Petals 4. Stamens ∞, adhering to the base of petals; filaments very short; anthers extrorse. Ovary free, terete, 1-celled, with 2 placentæ; stigmas 4. Fruit a capsule,

10-nerved, 2-valved; seeds on the margins of the valves.

1. E. CALIFORNICA Cham. Flowers orange—♃. Common. Summer.

### 5. Meconopsis Viguier.

Sepals 2, caducous. Petals 4. Stamens ∞; anthers laterally dehiscent. Ovary obovate, 1-celled, with more than 2 placentæ; style short, persistent; stigmas radiating. Fruit a 1-celled capsule, dehiscent by several slits near the apex.

1. M. HETEROPHYLLA Benth. Petals scarlet. ☉. Sunnyside. San Mateo. Niles. Summer.

### 6. Argemone Tourn. PRICKLY POPPY.

Sepals 2 or 3. Petals 4 or 6. Stamens ∞; anthers extrorse. Ovary ovate, 2-celled, with more than 2 placentæ; stigmas nearly sessile, radiating. Fruit a capsule, dehiscent by several valves near the apex.—☉.

1. A. HISPIDA Gray. Petals white.—Niles. Summer.

ORDER 2. POLYCARPICÆ. Floral parts with a tendency to arrange themselves spirally, and gradually merge into each other.

### Family 1. LAURINEÆ.

Perigonium 6-parted. Lobes 2-seriate, imbricate in æstivation. Disc adnate to the base of the perigonium; Stamens inserted on the margin of a disc, 6 or its multiple; anthers dehiscent by valves. Ovary 1-celled, compound of 3 carpidia, but only 1 ovule developed; style 1. Fruit baccate or drupaceous. Leaves alternate; stipules 0.

### 1. Oreodaphne Nees (*Umbellularia* Nutt).

Margin of perigonium deciduous. Stamens 9–12, in 3 rows; anthers of the two external rows introrse, of the third extrorse, their filaments 2-glandular at the base. Fruit a drupe resting on an enlarged thalamus.— ♄.

1. O. CALIFORNICA Nees. Common. Summer.

Bay tree. California laurel.

### Family 2. BERBERIDEÆ.

Sepals, petals and stamens 1–3-seriate, opposite each other. Anthers extrorse, dehiscent by valves. Ovary 1-celled, with ventral placenta ∞-ovulate. Fruit capsular, follicular or baccate, 1 to several-seeded. Leaves alternate; stipules 0.

### 1. Vancouveria Decaisne.

Sepals 6. Petals 6. Stamens 6. Fruit a follicle, several-seeded. Flowers white.— ♃.

1. V. HEXANDRA Morr. & Decaisne.—Marin County. Spring.

## 2. Berberis L.  Barberry.

Sepals 6, protected by bracts. Petals 6. Stamens 6. Fruit baccate, few-seeded. Flowers yellow.— ♄.

1. B. PINNATA Lag. Leaflets pinnately veined, shining above, acuminate; first pair of leaflets near the base of the petiole.—Colma. Mission Hills. Spring.

Berries glaucous, edible. Stem and root contain a yellow dye.

2. B. AQUIFOLIUM Pursh. Leaflets pinnately veined, shining above, acuminate; first pair of leaflets remote from the base of the petiole.— Crystal Springs. Spring.

3. B. REPENS Lindl. Leaflets pinnately veined, not shining above, not acuminate.— Marin County. Spring.

4. B. NERVOSA Pursh. Leaflets palmately nerved.—Marin County. Spring.

## Family 3. RANUNCULACEÆ.

Petals alternate with sepals and imbricate in æstivation. Stamens $\infty$; anthers extrorse (lateral), longitudinally dehiscent. Ovaries several, rarely reduced to 1, apocarpous. Fruit akene, follicle or berry. Stipules 0.

### 1. Pæonia Tourn.  Peony.

Sepals 5, unequal, persistent. Petals equal 5 or 10. Stamens $\infty$; ovaries few, seated on a fleshy disc, ∞-ovulate; stigmas sessile. Fruit of few follicles, coriaceous, ∞-seeded.

1. P. BROWNII Dougl. Petals dark red.—♃. Marin County. Summer.

## 2. Actæa L. BANEBERRY.

Sepals 4, equal, caducous. Petals 4. Ovary but 1, 1-celled with ventral placenta, ovules 2-seriate; stigma sessile. Fruit baccate, ∞-seeded.—♃. Flowers white.

1. A. SPICATA L. Berry red.—Redwood Peak. Tamalpais. Nicasio. Spring.

Narcotic poison.

## 3. Delphinium Tourn. LARKSPUR.

Sepals 5, petaloid; upper one calcarate. Corolla irregular; upper petal 0; 2 lateral petals projecting by their appendices into the calyx-spur. Ovaries not more than 5, ∞-ovulate. Fruit several follicles.

1. D. NUDICAULE Torr. & Gray. Leaves almost all radical, 3-5-lobed, segments broad, obtuse, mucronulate; flowers in lax racemes, scarlet.—☉. Contra Costa Range. Marin County. Summer.

2. D. CALIFORNICUM Torr. & Gray. Stem smooth, foliose; leaves 3-5-cleft, divisions lobed; flowers in strict, dense racemes; pedicels and flowers velvety, pale, sometimes tinged with violet.—♃. Saucelito. Mission Dolores. Summer.

3. D. decorum. Fisch. & Mey. Pubescent; lower leaves 5-lobed, sparingly toothed, upper leaves with narrow, linear divisions; flowers on long pedicels in an open raceme; follicles glabrous; flowers blue.—♃. Marin County. Summer.

4. D. variegatum Torr. & Gray. Pubescent; upper and lower leaves dissected, segments oblong or linear; flowers on long pedicels in an open raceme; follicles pubescent; flowers blue or white.—♃. Marin County. Crystal Springs. Summer.

5. D. simplex Dougl. Canescent; upper and lower leaves dissected, segments linear; flowers on short pedicels in dense racemes; follicles pubescent; flowers white, yellow, red, blue or dark purple.—⊙. Marin County. Summer.

This genus needs investigation in regard to its toxic properties. One of the species is known to the sheep-herders as dangerous to their flocks. Perhaps all the perennial species may contain an acrid, narcotic substance.

### 4. Aquilegia Tourn. Columbine.

Sepals 5, petaloid, equal. Petals 5, bilabiate, prolonged backward into a spur. Fruit 5 follicles.—♃.

1. A. truncata Fisch. & Mey. Flowers orange.—Common. Spring.

## 5. Ranunculus L. BUTTERCUP.

Sepals 5. Petals 5 or multiple, with a scale at the base. Ovaries $\infty$, 1-ovulate. Fruit $\infty$-akenes.

1. R. HEBECARPUS Hook. & Arn. Akenes hispid, roughened; floral parts of small size and reduced in their typical number.—⊙. Marin County. Niles. Spring.
Flaccid annual of the habit of Bowlesia.

2. R. MURICATUS L. Lower leaves rounded, 3-lobed, irregularly crenate; akenes tuberculate, spinose; floral parts typical and of ordinary proportions; petals yellow.—⊙⊙. Presidio. Summer.

3. R. MAXIMUS Greene. Leaves broad, alternately divided; radical leaves on very long petioles; leaflets laciniately lobed; akenes glabrous, flat, orbicular, thin-edged; styles long; petals more than 5, yellow.—♃. San Rafael. Corte Madera. Summer.

4. R. CALIFORNICUS Benth. Radical leaves pinnately ternate; leaflets laciniately cut; akenes glabrous, flat, firmly edged; styles short; petals more than 5; yellow.—♃. Common. Summer.

5. R. BLOOMERI Watson. Leaves 3-5-lobed, crenate; akenes lenticular; styles long, straight; petals yellow.—♃. Sunnyside. Summer.

6. R. FLAMMULA L. Leaves entire; akenes smooth; stamens more than 10; petals bright yellow.—♃. Marin County. Summer.

7. R. PUSILLUS Poir. Leaves entire; akenes granulate; stamens never more than 10; petals yellow.—☉. Inundated places near Napa. Summer.

8. R. LOBBII Hiern. Aquatic; leaves floating, 3-lobed; submerged leaves 0; styles long, filiform, with small terminal stigma; petals white with yellow claws.—♃. Olema. Summer.

9. R. AQUATILIS L. Aquatic with submerged capillary, multifid leaves; styles subulate, not longer than the ovary, introrsely stigmatose; petals white with yellow claws.—♃. Presidio. Marin County. Summer.

All perennial species possess an acrid, poisonous principle, which disappears by drying, so that the plant, avoided by animals in its growing state, becomes harmless when made into hay.

### 6. Myosurus L.  MOUSETAIL.

Sepals 5, calcarate. Petals 5, on filiform claws. Fruit ∞-akenes, crowded on an elongated, spike-like receptacle.—☉. Dwarf plants with radical leaves.

1. M. MINIMUS L.—Niles. Spring.

### 7. Thalictrum Tourn.  MEADOW-RUE.

Involucre 0. Sepals 4 or 5, very caducous. Petals 0. Ovaries ∞, 1-ovulate. Fruit ∞-capitate akenes. Leaves ternately compound. ♃.

1. T. FENDLERI Engelm. Flowers violet. Saucelito. San Mateo. Summer.

### 8. Anemone Hall.

Involucre 3-foliate. Sepals petaloid, 5 or a multiple. Petals 0. Ovaries ∞, 1-ovulate. Fruit ∞, capitate akenes.—♃.

1. A. GRAYI Kellogg & Behr. Flowers white, tinged with blue or purple.—Tamalpais. Camp Taylor. Spring.

### 9. Clematis L.  VIRGIN'S-BOWER.

Sepals 4, petaloid, not imbricate in æstivation. Petals 0. Ovaries 1-ovulate. Fruit ∞ akenes, caudate by the persistent style.

1. C. LASIANTHA Nutt. Tomentose; flowers solitary on long peduncles, pale.—♃. Niles. Summer.

2. C. LIGUSTICIFOLIA Nutt. Almost glabrous; Flowers paniculate, pale.—♃. San Mateo. Niles. Marin County. Summer.

ORDER 3. RANALES. Floral parts having a pronounced tendency to merge gradually into each other.

### Family 1. NYMPHÆACEÆ.

Sepals and petals inserted into a fleshy, cup-shaped thalamus. Stamens ∞, ∞-seriate, inserted with the petals; filaments flat, petaloid, anthers introrse, adnate. Carpidia several, immersed in the thalamus, and presenting the appearance of a ∞-celled ovary; ovules ∞, inserted on the partitions; styles united and forming a radiate, peltate stigma, projecting beyond the thalamus and persistent.

#### 1. Nuphar Smith. WATER-LILY.

Sepals more than 4, petaloid. Petals ∞, short, densely crowded round the thalamus. Stigma stipitate. Fruit a baccate pseudocarp. ♃. Aquatic. Flowers yellow.

1. N. POLYSEPALUM Engelm. Sepals more than 7.—Near the Marine Hospital. Taylorville, Paper-mill Creek. Summer.

SERIES 3. PERIGYNÆ. Thalamus cup-shaped, bearing sepals, petals and stamens, and receiving the carpidia in shape of a receptacle, which in many instances surrounds the carpidia enough to form an inferior ovary.

Section 1. CALYCIFLORÆ. Petals present and distinct, 4 or 5. Stamens 4 or 5, or their multiples.

ORDER 1. MYRTALES. Calyx lobes valvate in æstivation. Petals as many, or a mul-ple, rarely a divisor. Carpidia united into a single ovary.

### Family 1. HALORRHAGIDEÆ.

Calyx tube adnate to the ovary; limb 4-parted. Petals, if present, alternate with calyx lobes. Stamens not more than twice as many as petals; if of equal number opposite to them. Placentæ central, 1-ovulate. Styles as many as cells.

#### 1. **Hippuris** L.   Mare's Tail.

Flowers ☿, or sometimes polygamous, monandrous; filament subulate. Ovary 1-celled, 1-ovulate, style filiform. Fruit a caryopsis.—♃. Aquatic. Leaves verticillate; flowers minute, solitary in the axils.

1. H. VULGARIS L.—Marine Hospital (at present extinct).   Summer.

#### 2. **Myriophyllum** L.   Water-Milfoil.

Monœcious. ♂: petals caducous; stamens 8. ♀: calyx limb and petals minute; stigma villous; ovary 4-celled, 4-ovulate.

1. M. SPICATUM L.—Marine Hospital. Camp Taylor.   Summer.

### Family 2. ONAGRACEÆ.

Calyx tube adnate to the ovary. Petals alternate with them, not valvate in æstivation. Stamens inserted with petals, as many or a multiple; anthers introrse. Placentæ central, ∞-ovulate. Style 1. Stipules 0.

### 1. Jussiæa L.

Calyx tube not prolonged beyond the ovary, but forming an epigynous disc; lobes 4-6, persistent. Petals 4-6; stamens twice as many. Cells of ovary as many as calyx-lobes; stigma capitate. Fruit a septicidal capsule.

1. J. REPENS L. Flowers yellow; leaves alternate.—♃. Aquatic. Niles. Alvarado. Summer.

### 2. Zauschneria Presl.

Calyx-tube prolonged considerably above the ovary, funnel-shaped, colored, with 4-lobed, deciduous limb. Petals not exceeding the calyx-lobes, 2-cleft, erect. Stamens 8, included. Ovary 4-celled, 4-valved. Seeds with a hairy crown. Fruit a capsule, linear, imperfectly 4-celled, ∞-seeded. Flowers scarlet; upper leaves alternate.—♃.

1. Z. CALIFORNICA Presl. — Oakland Waterworks. Niles. Summer.

### 3. Epilobium L. WILLOW-HERB.

Calyx-tube prolonged but little beyond the ovary; limb 4-parted, soon deciduous. Petals 4, obovate, or obcordate, inserted into an annular disc on the summit of the calyx-tube. Stamens 8. Ovary 4-celled. Fruit a 4-angular,

4-celled, loculicidal capsule. Seeds with a crown of long hair.

1. E. FRANCISCANUM Barbey. Stem angular; leaves serrulate; stamens shorter than the style; petals purple.—♃. Fort Point. Summer.

2. E. MINUTUM Lindl. Stem terete; leaves almost entire; four of the stamens as long as the style; petals rose-color.—☉. Tamalpais. Summer.

### 4. Gayophytum Juss.

Calyx tube not prolonged beyond the ovary, limb 4-parted, reflexed, deciduous. Petals 4. Stamens 8; anthers versatile, those opposite the petals smaller and usually sterile. Ovary 2-celled; style short; stigma capitate or clavate. Fruit a capsule, 2-celled, uneqally 4-valved.—☉. Slender herbs.

1. G. DIFFUSUM Torr. & Gray. Petals pale rose-color.—Saucelito. Spring.

### 5. Œnothera L. EVENING PRIMROSE.

Calyx tube more or less prolonged above the ovary; lobes reflexed. Petals 4. Stamens 8; anthers versatile. Ovary 4-celled; ovules ∞; style filiform. Fruit a 4-celled capsule, loculicidal; seeds ∞, smooth. Leaves alternate.

1. Œ. BIENNIS L. Stigma lobes linear; flowers in a leafy spike, yellow.—☉ ☉. San Mateo. Niles. Summer.

2. Œ. CALIFORNICA Watson. Stigma lobes linear; flowers axillary, white, at last pinkish. ♃. Niles. Summer.

3. Œ. OVATA Nutt. Stigma capitate; calyx tube filiform, above the ovary; petals yellow. ♃. Common. Spring.

Leaves and flowers formerly in use as a salad.

4. Œ. CHEIRANTHIFOLIA Hornem. Stigma capitate; calyx tube obconic; capsule 4-angular with acute angles; canescent; leaves thick, entire; petals yellow.—♃. Seashore. Summer.

5. Œ. MICRANTHA Hornem. Stigma capitate; calyx tube obconic; capsule 4-angular with acute angles, very much contorted; leaves denticulate; petals yellow.—⊙. Niles. Alvarado. Summer.

6. Œ. DENTATA Cav. Stigma capitate; calyx tube obconic; capsule elongated, linear, shortly beaked; petals yellow.—⊙. Common. Summer.

7. Œ. STRIGULOSA Torr. & Gray. Stigma capitate; calyx tube obconic; capsule elongated, linear, obtuse, petals yellow.—⊙. Common. Summer.

### 6. Godetia Spach.

Calyx-tube prolonged beyond the ovary, funnel-shaped, the lobes reflexed. Petals 4. Stamens 8, unequal; anthers basifixed. Capsule ovate to linear; seeds smooth, ∞.—⊙.

1. G. PURPUREA Watson. Flowers in a leafy, terminal cluster; capsule oblong; seeds in a double row in each cell; petals purple. Common. Summer.

2. G. LEPIDA Lindl. Flowers in short simple spike; capsule oblong, ovate; seeds in a simple row in each cell; petals rose-colored, with a darker spot near the apex.—Niles. Summer.

3. G. QUADRIVULNERA Spach. Capsule linear, sessile, villous, 2-costate; petals purple. Common. Summer.

4. G. TENELLA Watson. Capsule linear, sessile, puberulent, not costate; petals purple. Niles. Summer.

5. G. AMŒNA Lilja. Capsule linear, attenuate at each end, pedicillate; plant minutely puberulent; petals rose-color, tinged with purple.—Common. Summer.

6. G. EPILOBIOIDES Watson. Capsule linear, attenuate into a short pedicel, acuminate at the apex; plant tomentosely pubescent; petals rose-color.—Niles. Summer.

### 7. Clarkia Pursh.

Calyx-tube prolonged beyond the ovary, obconical, deciduous, the lobes reflexed. Petals 4, unguiculate. Stamens 8, alternately depauperate; anthers basifixed. Capsule linear; seeds $\infty$, smooth.—⊙.

1. C. ELEGANS Dougl. Petals rose-color. Alameda. Summer.

### 8. Eucharidium Fisch. & Mey.

Calyx-tube linear, elongated beyond the ovary, its lobes reflexed, deciduous. Petals 4, unguiculate. Stamens 4, alternate with the petals; anthers basifixed. Seeds ∞, smooth. ☉.

1. E. CONCINNUM Fisch. & Mey. Flowers red.—Marin County. Summer.

### 9. Boisduvalia Spach.

Calyx-tube prolonged beyond the ovary; funnel-shaped, deciduous; the lobes erect. Petals 4, sessile, 2-lobed. Stamens 8, those opposite the petals shorter; anthers basifixed. Capsules 4-celled; seeds ∞, smooth.—☉. Flowers purple.

1. B. DENSIFLORA Watson. Canescent; capsules oblong; partitions of the dehiscent capsule separating from the valves.—Common. Summer.

2. B. GLABELLA Walpers. Capsules oblong, almost straight; partitions of the dehiscent capsule adhering to the valves.—Common. Summer.

3. B. CLEISTOGAMA Curran. Capsules oblong, 4-angular, curving outward from the stem; partitions of the dehiscent capsule adherent to the valves.—Suñol. Summer.

## Family 3. LYTHRARIÆ.

Calyx tube not adnate to the ovary. Stamens inserted with the petals, and of definite number; anthers introrse Ovary free, with central placentæ, compound of carpidia; ovules $\infty$; style 1.

### 1. Lythrum L.    LOOSE-STRIFE.

Calyx tubular, 8–12-costate; limb 8–12-dentate; dentitions alternating in size. Petals 4–6, opposite the smaller dentitions of the calyx. Stamens 8–12. Ovary 2-celled, covered by the persistent calyx. Flowers purple.

1. L. ALATUM Pursh.—♃. Tamalpais. Summer.

ORDER 2. ROSIFLORÆ. Calyx lobes imbricate in æstivation, as many as the perigynous petals. Ovaries several, in different degrees of consolidation, in rare cases reduced to a single one, the eccentric position of which indicates the abortion of its fellows.

### Family 1. AMYGDALACEÆ.

Calyx free, 5-cleft, deciduous. Petals 5, convolute in æstivation. Ovaries almost always reduced to one, 1-celled, 2-ovulate. Fruit a drupe, by abortion generally 1-seeded. Leaves alternate, stipulate; stipules caducous.

### 1. Prunus L.  PLUM. CHERRY.

Drupe fleshy; endocarp not rugose.— ♄.

1. EMARGINATA Walpers. Flowers in a corymb; petals white.—Tamalpais. Spring.

Black, cherry-like fruit. Not poisonous, although not edible.

2. P. DEMISSA Walpers. Flowers in terminal racemes; petals white.—Mills Seminary. Marin County. Spring.

Dark red or purplish cherry. Edible. "Choke-cherry."

3. P. ILICIFOLIA Walpers. Flowers in leafless racemes from the axils of evergreen leaves; petals white. — Common. Spring. "Holly-leaf Cherry."

Red or purplish cherry, not of bad taste, but suspicious. The leaves at times prove poisonous to sheep and cattle, probably only when by their withering hydrocyanic acid is developed, the smell of it in the withering leaves becomes very perceptible.

### 2. Nuttallia Torr. & Gray.

Flowers polygamous. Ovaries 5. Fruit several 1-seeded drupes. Flowers white.— ♄.

1. N. CERASIFORMIS Torr. & Gray. Flowers white.—Common. Spring.

Drupes blue, of agreeable but slightly bitter taste, but suspicious.

## Family 2. SPIRÆEÆ.

Ovaries verticillate, rarely reduced to 1. Fruit follicular.

### 1. Spiræa L.

Seeds small exalbuminous, with membranaceous testa.

1. S. DISCOLOR Pursh. Flowers reddish.— ♄. Common. Summer.

### 2. Neillia Don. NINE-BARK. BRIDAL-WREATH.

Seeds albuminous, with a shining crustaceous testa. Flowers white.— ♄.

1. N. OPULIFOLIA Benth. & Hook.—Common. Summer.

## Family 3. DRYADACEÆ.

Ovaries collected around a convex receptacle, rarely reduced to 1, 1-ovulate. Fruit indehiscent, frequently drupaceous.

### 1. Rubus L. BLACKBERRY. RASPBERRY.

Calyx persistent, concave or flattened; limb 5-parted. Stamens ∞. Ovaries on a convex receptacle. Fruit drupes ∞, on a conical receptacle.

1. R. NUTKANUS Moc. Leaves palmately lobed; petals white.— ♄. Common. Summer. "Salmon-berry."

2. R. SPECTABILIS Pursh. Armed with stout prickles; leaves 3-foliolate, underneath slightly

pubescent; flowers almost solitary, red. — ♄. Colma. Saucelito. Summer.

Fruit looks like a raspberry, red or yellow, rather insipid.

3. R. LEUCODERMIS Dougl. Armed with stout prickles; leaves 3-5-foliolate, leaflets underneath white-tomentose; flowers in few-flowered corymbs, white.— ♄. Los Gatos Creek. Summer.

Fruit a raspberry, red, of agreeable flavor.

4. R. URSINUS Cham. & Schlecht. Armed with slender bristles or bristle-like prickles; leaves generally 3-foliolate, not tomentose underneath; flowers white.— ♄. Common. Summer. "Wild-blackberry."

### 2. Fragaria L. STRAWBERRY.

Calyx concave or flattened, persistent, 5-lobed, augmented by 5 bractlets. Petals 5. Stamens a multiple of 5. Ovaries ∞, on a convex receptacle; styles lateral. Fruit ∞ akenes on an enlarged, fleshy receptacle.—♃.

1. F. CHILENSIS Ehr. Leaves perfectly smooth and shining on their upper surface. Flowers white.—Presidio. Cliff House. Point Bonita. Summer.

2. F. CALIFORNICA Cham. & Schlecht. Leaves sparingly villous and not shining on their upper surface.—Common. Summer.

### 3. Potentilla L. FIVE-FINGER.

Calyx concave or flattened, 5-lobed, augmented by 5 bractlets. Petals 5. Stamens a multiple of 5. Ovaries ∞, on a slightly conical receptacle. Styles lateral. Fruit ∞ akenes. Receptacle not enlarged.

1. P. GLANDULOSA Lindl. Flowers cymose; petals pale yellow, sometimes white.—♃. Contra Costa range. Summer.

2. P. ANSERINA L. Flowers axillary, solitary, yellow.—♃. Common on moist places. Summer.

### 4. Horkelia Cham. & Schlecht.

Calyx campanulate, limb 5-parted, augmented by 5 bractlets. Petals 5. Stamens a multi- of 5. Ovaries inserted on a conical receptacle; styles subterminal. Fruit akenes.—♃.

1. H. CALIFORNICA Cham. & Schlecht. Glandular pubescent; calyx-tube deeply campanulate; petals white.—Common. Summer.

2. H. KELLOGGII Greene. Silky pubescent; calyx-tube spreading; cupuliform; petals white. Alameda. Summer.

### Family 4. SANGUISORBACEÆ.

Calyx-tube contracted at its apex, including the ovaries. Petals often 0, Stamens and ovaries often reduced in number. Ovaries 1-ovulate.

### 1. Adenostoma Hook. & Arn.
#### Chemisal. Greasewood.

Calyx funnel-shaped; tube 10-costate; limb 5-parted. Petals 5; stamens a multiple. Ovary 1, with truncate, pubescent apex. Fruit an akene, included in the persistent calyx-tube. Flowers white.— ♄.

1. A. fasciculatum Hook. & Arn.—Common. Summer.

### 2. Alchemilla Tourn. Lady's Mantle.

Calyx-tube urceolate, persistent; limb 4-parted, with 4 minute, deciduous bractlets. Petals 0. Stamens reduced, 1–4. Ovaries 1–4; styles basilar. Fruit 1–4 akenes.

1. A. arvensis Scop.—⊙. Common. Spring.

### 3. Acæna Vahl.

Calyx-tube contracted at the throat, oblong, angular, the angles armed with glochidiate prickles; limb persistent. Petals 0. Stamens reduced 1–5. Ovaries 1 or 2; styles terminal; stigmas penicillate. Fruit an akene, enclosed in the persistent calyx.— ♃.

1. A. trifida Ruiz & Pavon.—Common. Summer.

### 4. Cercocarpus HBK. Mountain Mahogany.

Calyx-tube cylindrical, persistent; limb turbinate, 5-lobed, deciduous. Petals 0. Stamens multiple of 5. Ovary 1; style terminal, long exserted. Fruit an akene, linear, terete, caudate by the elongated, plumose, twisted style.— ♄

1. C. PARVIFOLIUS Nutt.—San Leandro. Lake Chabot. Spring.

### Family 5. ROSACEÆ.

Carpidia ∞, 1-ovulate, indehiscent, included in the fleshy tube of the calyx. Stamens ∞.

#### 1. Rosa Tourn. ROSE.

Calyx-tube urceolate, constricted at the throat; limb 5-parted. Petals 5. Styles lateral, exserted. Fruit ∞-akenes, included in the fleshy calyx-tube.— ♄ .

1. R. CALIFORNICA Cham. & Schlecht. Foliage and inflorescence tomentose; fruit globose; calyx lobes persistent.—Common. Summer.

2. R. GYMNOCARPA Nutt. Glabrous; petioles and stipules glandular; fruit ovate; calyx lobes deciduous.—Tamalpais. Summer.

### Family 6. CALYCANTHEÆ.

Calyx fleshy; tube obconical; limb divided into ∞-seriate lobes. Petals 0. Stamens ∞; anthers extrorse. Ovaries ∞, 1-ovulate inserted on the whole inner surface of the calyx-tube; styles ∞, terminal, exserted. Fruit ∞-akenes, enclosed in the fleshy calyx-tube. Leaves opposite. Stipules 0.

### 1. Calycanthus Lindl. Spice-Bush.

Inner lobes of the ∞-seriate calyx-limb smaller than the external. Outer series of stamens fertile, inner sterile. Flowers dark purple.— ♄.

1. C. OCCIDENTALIS Hook. & Arn.—Sonoma. Napa. Summer. Leaves and branches fragrant when bruised.

### Family 7. POMACEÆ.

Calyx-tube adnate to the ovary; limb 5-parted. Carpidia 2–5, 2-ovulate; styles 2–5. Fruit baccate.

#### 1. Photinia Lindl.
(*Heteromeles* J. Rœmer.) Toyon.

Carpidia 2, imperfectly united and only half immersed in the calyx-tube; styles 2. Calyx-limb growing fleshy in fruit and covering the upper half of the carpidia, which are 1-seeded by abortion. Flowers in corymbose panicles. ♄.

1. P. ARBUTIFOLIA Rœmer. Flowers white. Common. Summer.

Berries coral red, of astringent taste, but used by the original settlers in the preparation of a refreshing drink.

#### 2. Amelanchier Medik.
June-berry. Service-berry.

Ovules of carpidion separated by an incomplete septum, 1 ovule becoming abortive in the ripe fruit.— ♄. Flowers in racemes.

1. A. ALNIFOLIA Nutt. Flowers white. Common. Summer. Berries purplish, edible.

ORDER 3. LEGUMINOSÆ. Ovary 1, 1-celled. Fruit a legume. Leaves alternate, stipulate.

### Family 1. PAPILIONACEÆ.

Corolla papilionaceous. Stamens 10.

#### 1. Pickeringia Nutt.

Petals of the carina not connate. Stamens not diadelphous. Legume linear, compressed. Flowers purple; leaves palmately 3-foliolate, evergreen; stipules evanescent.— ♄.

1. P. MONTANA Nutt.—Tamalpais. Crystal Springs. Wright's Station. Summer.

#### 2. Thermopsis R. Brown.

Petals of the carina partly connate. Stamens not diadelphous. Legume linear, compressed. Flowers yellow; leaves palmately 3-foliolate; stipules persistent, foliaceous.— ♃.

1. T. CALIFORNICA Watson.—San Rafael. Saucelito. Summer.

#### 3. Lupinus L.   LUPINE.

Stamens monadelphous; anthers alternately oblong and reniform. Leaves palmate; stipules adnate.

1. L. ARBOREUS Sims. Slightly silky pubescent, pubescence appressed; flowers on slender pedicels; bracts linear, equaling the calyx, deciduous; superior and inferior calyx lip nearly equal; ovules more than 6 in the legume; petioles short; flowers yellow.—⊙ ⊙. Common. Summer.

2. L. CHAMISSONIS Esch. Appressed silky pubescence; pubescence dense; flowers on slender pedicels; bracts lanceolate, shorter than the calyx, deciduous; superior and inferior calyx lip nearly equal; ovules more than 6; petioles short; flowers violet or pale.— ♄. Common. Summer.

3. L. DOUGLASII Agardh. Almost tomentose; flowers on slender pedicels; bracts linear exceeding the calyx, deciduous; superior and inferior calyx lip nearly equal; ovules more than 6; petioles short; flowers purple.—♃. Common. Summer.

4. L. POLYPHYLLUS Lindl. Leaflets glabrous above; flowers on long pedicels; bracts lanceolate, not longer than the calyx; superior and inferior calyx lip nearly equal, entire; ovules more than 6; petioles considerably longer than the leaflets; flowers blue, purple or white.—♃. Common. Summer.

5. L. RIVULARIS Dougl. Leaflets glabrous above; flowers on long pedicels; bracts setaceous, exceeding the calyx; superior and inferior

calyx-lip nearly equal; ovules more than 6; petioles equalling the leaflets; flowers purple. ♃. Tamalpais. Summer.

6. L. LITTORALIS Dougl. Leaflets glabrous above; flowers on short pedicels. Racemes short; bracts setaceous, exceeding the calyx; ovules more than 6; petioles longer than the leaflets; flowers violet marked with yellow.—♃. Marin County near the seacoast. Summer.

7. L. ALBICAULIS Dougl. Bracts subulate, shorter than the calyx; deciduous; calyx-lips nearly of equal length, the upper narrowed and dentate; vexillum acute, its margins near the apex coherent; ovules about 6; petioles not longer than the leaflets; flowers pale blue. ♃. Marin County. Summer.

8. L. AFFINIS Agard. Pubescent; racemes formed by regular whorls; bracts short; superior calyx-lip 2-cleft; legume linear; ovules about 6; petioles twice as long as the leaflets; flowers blue with a white spot on the vexillum, which afterwards turns red. — ☉. Common. Spring.

9. L. NANUS Dougl. Pubescent; racemes formed by regular whorls; bracts exceeding the calyx; superior calyx-lip 2-cleft; ovules 6-8; petioles much longer than the leaflets; flowers blue marked with white.—☉. Berkeley. Spring.

10. L. micranthus Dougl. Villous; racemes formed by regular whorls; bracts shorter than the calyx; superior calyx-lip 2-lobed; lobes short 3-angular; lower lip almost entire; flowers blue, sometimes marked with white.—⊙. Common. Spring.

11. L. trifidus Watson. Villous; racemes generally reduced to a single whorl; superior calyx-lip 2-lobed, inferior calyx-lip 3-cleft; flowers blue marked with white.—⊙. San Francisco sand hills. Spring.

12. L. truncatus Nutt. Nearly glabrous; flowers of the raceme scattered, not in whorls; superior calyx-lip 2-cleft; leaflets truncate at the apex; legume 8-ovuled; petals dark purple.—⊙. Contra Costa hills. Spring.

13. L. microcarpus Sims. Villous; calyx densely villous, its lips dentate, the upper very short; legume ovate, 2-ovuled; flowers purple, rose-color or white.—⊙. Common. Spring.

14. L. densiflorus Benth Villous; calyx almost glabrous; legume ovate, 2-ovuled; flowers yellow or pale.—⊙. Common. Spring.

### 4. Trifolium Tourn.  Clover.

Diadelphous. Petals persistent and adnate to the stamineal tube. Legume utriculate, irregularly dehiscent. Leaves palmately compound; stipules adnate.

1. T. REPENS L. All peduncles axillary, longer than the leaf. Petals white.—♃. Common. Summer.

2. T. MACRÆI Hook. & Arn. Heads not involucrate; flowers sessile; petals dark purple. ⊙. Marin County. Spring.

3. T. CILIATUM Nutt· Heads not involucrate; flowers shortly pedicillate; calyx teeth lanceolate, rigid, ciliate at the bottom; petals white or purplish.—⊙. Common. Spring.

4. T. GRACILENTUM Torr. & Gray. Heads not involucrate; flowers shortly pedicillate; leaflets obovate; petal rose-color.—Common. Spring.

5. T. BIFIDUM Gray. Heads not involucrate; flowers shortly pedicillate; leaflets bifid at the apex; corolla rose-color.—⊙. Common. Spring.

6. T. INVOLUCRATUM Willd. Heads involucrate; involucre herbaceous, deeply lobed, the lobes deeply dentate; corolla not becoming inflated; legume more than 4-ovulate; petals rose-color.—⊙. Common. Spring.

7. T. TRIDENTATUM Lindl. Heads involucrate; involucre herbaceous, deeply lobed, the lobes deeply dentate; corolla not becoming inflated; legume 1-2-ovulate; calyx teeth rigid, abruptly narrowed into a spinulose apex; petals considerably exceeding the calyx, purple with paler or white tips.—⊙. Common. Spring

8. T. pauciflorum Nutt. Heads involucrate; involucre herbaceous, deeply lobed, the lobes deeply dentate; corolla not becoming inflated; legume 1-2-ovulate; petals scarcely exceeding the calyx, purple or rose-color.—☉. Common. Spring.

9. T. microcephalum Pursh. Heads involucrate; involucre membranaceous, its lobes entire; corolla not becoming inflated; petals pale.—☉. Common. Spring.

10. T. microdon Hook. & Arn. Heads involucrate; involucre membranaceous, its lobes 3-dentate; corolla not becoming inflated; petals white or pale.—☉. Common. Spring.

11. T. barbigerum Torr. Heads involucrate; involucre broad, ∞-dentate; corolla becoming inflated; petals purple.—☉. Common. Spring.

12. T. fucatum Lindl. Heads involucrate; involucre deeply lobed; corolla becoming inflated, pale.—☉. Common. Spring.

13. T. depauperatum Desv. Involucre almost 0, reduced to a mere disc; corolla becoming inflated, white or purple.—☉. Common. Spring.

### 5. **Melilotus** Tourn.   Melilot.

Diadelphous. Petals free from the column of stamens and deciduous. Legume globose. Flowers racemose. Leaves pinnately 3-foliolate; stipules adnate.

1. M. PARVIFLORA Desf. Flowers yellow. ⊙. Common the year round.

Native of the Mediterranean region, but at present perfectly naturalized.

### 6. Medicago Tourn. MEDICK.

Diadelphous. Petals free from the column of stamens and deciduous. Legume globose. Flowers racemose. Leaves pinnately 3-foliolate; stipules adnate.

1. M. SATIVA L. Legume without prickles, spirally rolled up, the spiral leaving an open space in the center; flowers purple. — ♃. Common all the year round. "Alfalfa."

Native of Europe, but escaped from cultivation in different countries.

2. M. DENTICULATA Willd. Legume spirally rolled up and armed with a double row of prickles; flowers yellow.—⊙. Common all the year round.

Native of Europe, a fodder plant, but owing to the prickles that infest its legume, not desirable in wool-growing districts.

3. M. LUPULINA L. Legume without prickles, spirally rolled up, the spiral leaving no open space in the center; flowers yellow.—⊙. Common all the year round.

Native of Europe, now naturalized.

## 7. Hosackia Dougl.

Diadelphous. Petals unguiculate. Legume cylindrical, linear, almost straight, dehiscent, sessile. Ovules more than 2. Flowers umbellate, or by depauperation solitary. Leaves imparipinnate, 2 to several-foliolate; stipules often minute.

1. H. STIPULARIS Benth. Stipules large, foliaceous; flowers purplish.—♃. Contra Costa range. Summer.

2. H. GRACILIS Benth. Stipules scarious; vexillum yellow; wings and carina purple.—♃. Colma. Talmalpais. Summer.

3. H. STRIGOSA Nutt. Stipules reduced to blackish glands; rachis of leaf flattened; vexillum attenuate below, carina obtuse; flowers yellow.—⊙. Common. Spring.

4. H. PARVIFLORA Benth. Stipules reduced to blackish glands; rachis of leaf flattened; vexillum cordate; carina acute; flowers pale.—⊙. Common. Spring.

5. H. PURSHIANA Benth. Stipules reduced to glands; leaves almost sessile; flowers peduncled, pale; vexillum pink.—⊙. Common. Summer.

6. H. SUBPINNATA Torr. & Gray. Stipules gland-like, leaves petioled; flowers almost sessile; legume 5-seeded; petals yellow.—⊙. Common. Summer.

7. **H. brachycarpa** Benth. Stipules gland-like; leaves petioled; flowers almost sessile; legume less than 5-seeded; petals yellow.—⊙. Common. Summer.

### 8. Syrmatium Vogel. (*Hosackia,* Dougl.)

Diadelphous. Petals unguiculate. Legume indehiscent, incurved. Ovules less than 3. Umbels few-flowered. Stipules reduced to dark colored glands.

1. **S. glabrum** Torr. Glabrous; calyx teeth narrow, erect; flowers yellow.—♃. Common. Summer.

2. **S. cytisoides** Benth. Glabrous. Calyx teeth attenuate, recurved; flowers yellow.—♃. Half Moon Bay. Crystal Springs. Summer.

3. **S. tomentosum** Hook. & Arn. Pubescent throughout; calyx teeth filiform; flowers yellow.—♃. Marine Hospital. Summer.

### 9. Psoralea L.

Diadelphous. Glandular. Wings united to the carina. Ovary sessile, 1-ovulate. Legume included in the calyx, indehiscent. Leaves imparipinnate, 3–5-foliolate; stipules not adnate.

1. **P. orbicularis** Lindl. Stems prostrate; corolla purple, sometimes white.—♃. Common. Summer.

2. P. strobilina Hook. & Arn. Stem erect; stipules large, membranaceous, acuminate; peduncles shorter than the leaves; flowers in short spikes; bracts large, deciduous; corolla purple.—♃. Contra Costa range. Summer.

3. P. macrostachya DC. Stem erect; stipules small, lanceolate; peduncles much exceeding the leaves; spikes cylindrical; bracts large, acuminate; corolla purple.—♃. Marin County. Summer.

4. P. physodes Dougl. Stems erect; stipules small, lanceolate; peduncles about equaling the leaves; flowers in racemes; bracts small; corolla white or purplish.—♃. Coast Range. Summer.

### 10. Amorpha L. False Indigo.

Monadelphous. Vexillum unguiculate. Wings and carina 0. Ovary sessile, 2-ovulate. Legume very late, dehiscent. Glandular. Leaves imparipinnate.—♄.

1. A. californica Nutt. Flowers purple. Marine County. Summer.

### 11. Glycyrrhiza L. Liquorice.

Diadelphous. Calyx not bracteolate. Vexillum, wings and carina straight. Anthers confluent. Legume ovate, compressed, few-seeded, echinate. Glandular. Leaves imparipinnate.—♃.

1. G. LEPIDOTA Nutt. Flowers ochroleucous.—Niles. Summer,

## 12. Astragalus L.
RATTLE-WEED. LOCO-WEED.

Diadelphous. Carina blunt. Legume more or less divided by the intrusion of the dorsal suture. Seeds reniform. Leaves imparipinnate.

1. A. DIDYMOCARPUS Hook. & Arn. Legume transversely wrinkled, 2-seeded; corolla white and violet.—Contra Costa. Summer.

2. A. TENER Gray. Legume ∞-seeded, not inflated; corolla white, sometimes violet tipped. ☉.—Golden Gate Park. Cemetery. Summer.

3. A. LEUCOPHYLLUS Torr. & Gray. Legume inflated, stipitate in the calyx; corolla pale. ♃.—Suñol. Summer.

4. A. CROTALARIÆ Gray. Legume inflated, sessile in the calyx; stipules distinct; corolla white, more than twice the length of the calyx. ♃.—Sandhills near San Francisco. Summer.

5. A. MENZIESII Gray. Legume inflated, sessile in the calyx; stipules connate, opposite the petiole; corolla white, more than twice the length of the calyx.—♃.

6. A. DOUGLASII Gray. Legume inflated, sessile in the calyx; corolla pale, hardly twice the length of the calyx.—♃. Coast Range. Summer.

7. A. PYCNOSTACHYUS Gray. Legume coriaceous, not inflated, lenticular, reticulated; ovules 5; seeds less than 4; corolla pale.—♃. Bolinas Bay. Summer.

Some species of Astragalus have the reputation of being poisonous to cattle, sheep and horses. The specimens sent to the Academy of Sciences do not belong to any species found in the territory of the local flora. The whole matter is still wrapt in a mystery, like the one connected with the properties of *Gastrolobium*, a leguminous plant in Australia.

### 13. Vicia L. VETCH.

Diadelphous. Style filiform; apex pilose. Leaves paripinnate, terminating in branched tendrils.

1. V. GIGANTEA Hook. Flowers in a raceme; leaflets more than 9 pairs; corolla pale-purple.—♃. Coast Range. Summer.

2. V. AMERICANA Muhl. Flowers in racemes; leaflets less than 9 pairs; corolla purple.—♃. Common. Summer.

3. V. EXIGUA Nutt. Flowers pedunculate, solitary, rarely 2; leaflets less than 5 pairs; corolla purple.—☉. Contra Costa. Spring.

4. V. SATIVA L. Flowers almost sessile; leaflets more than 4 pairs; corolla violet.—☉. Fields, hedges. Spring.

Escaped from cultivation, originally European.

### 14. Lathyrus L. WILD PEA.

Style ventrally flattened toward the apex, concave, pilose along the inner side. Leaves paripinnate, ending in branched tendrils.—♃.

1. L. VESTITUS Nutt. Leaves cirrhate; peduncles ∞-flowered; petals rose-color, changing before withering into violet.—♃. Common. Summer.

2. L. PALUSTRIS L. Leaves cirrhate; peduncles 2-6-flowered; petals purple.—Marin County. Summer.

2. L. LITTORALIS Endl. Leaves not cirrhate; leaflets 1-3 pairs and a small terminal one; vexillum purple; wings and carina pale. Marin County. Summer.

### 15. Cercis L.
RED-BUD. JUDAS-TREE.

Stamens not diadelphous, anterior longer than the posterior ones; anthers versatile. Petals unguiculate, those forming the carina separate, and larger than the rest. Legume flat, stipitate, ∞-seeded; ventral suture winged.— ♄. Flowers purple in axillary fascicles. Leaves simple, appearing after the flowers.

1. C. OCCIDENTALIS Torr. Petals rose-color. Suñol. Spring.

Order 4. **DAPHNALES.** Petals 0. Perigonium and stamens inserted in a perigynous disc. Stamens 2 or its multiple. Carpidion 1, ovules never basilar. Stipules 0.

### Family 1. THYMELACEÆ.

Perigonium gamosepalous. Ovary 1-celled, 1-ovulate; ovule lateral, attached near the apex; style lateral.

#### 1. Dirca L.    Leatherwood.

Flowers ☿. Perigonium corolline, campanulate; limb obliquely truncate. Stamens 8, the alternate ones shorter. Hypogynous scales 0. Style filiform, subterminal. Fruit a drupe. ♄.

1. D. occidentalis Gray. Perigonium greenish-yellow; drupe orange.—Contra Costa hills. Coast Range. Spring.

Bark acrid, blistering when applied to the skin.

### Family 2. LORANTHACEÆ.

Evergreen, dichotomous, parasitic perennials. Branches articulate. Leaves opposite; stipules 0. Flowers epigynous. Stamens inserted on the perigonium, opposite its lobes and of the same number. Ovary 1-celled, 1-ovuled. Fruit baccate. Seed often containing several embryos.

[Affinities doubtful, perhaps connecting the Gymnosperm families of *Taxineæ* and *Gnetaceæ* to the Angiosperm families mentioned here and the *Proteaceæ*.]

### 1. Phoradendron Nutt.  MISTLETOE.

Flowers diclinous; ♂ and ♀, on separate spikes. Anthers 2-celled. Stigma sessile. Fruit crowned by the persistent perigonium.

1. P. FLAVESCENS Nutt.— ♄. Ross Station. Spring.

Parasite on oaks, poplars, etc.

### 2. Arceuthobium Bieb.  PINE MISTLETOE.

Diœcious. ♂ flowers sessile. Lobes of the perigonium ovate, spreading. Anthers sessile, inserted on the middle of the perigonium lobes, 1-celled, transversely dehiscent. ♀ flowers shortly pedicillate, compressed, with 2-dentate limb. Stigma sessile. Berry transversely dehiscent at the base.

1. A. OCCIDENTALE Engelm.— ♄. Contra Costa mountains. Autumn.

Parasite on conifers.

### Series 4. MONOCHLAMYDEÆ.

Flowers incomplete. Number of carpidia corresponding to number of floral parts.

ORDER 1. SERPENTARIÆ. Leafy plants containing chlorophyll, as distinguished from the other order of the series, the *Rhizantheæ*, which are parasitic plants without chlorophyll.

### Family 1. ARISTOLOCHIEÆ.

Perigonium gamosepalous. Stamens 6, or a multiple. Ovary inferior, 6-celled.

#### 1. **Asarum** Tourn.  WILD GINGER.

Stamens 12, f r e e . Perigonium entirely epigynous; limb regular. Rhizome dichotomous.—♃.

1. A. CAUDATUM Lindl. Flower dark-purple.—Strawberry Valley. Tamalpais. Spring.

#### 2. **Aristolochia** Tourn.  DUTCHMAN'S PIPE.

Stamens 6, gynandrous. Limb of perigonium oblique, irregularly cleft.

1. A. CALIFORNICA Torr. Flowers greenish; twining creeper.—♃. Saucelito. Ocean lake. Ross Station. Spring.

SERIES 5. J U L I F L O R Æ. Inflorescence dense; flowers apetalous. Leaves never compound.

ORDER 1. Flowers d i c l i n o u s; ♂ flowers amentaceous. Fruit 1-seeded, indehiscent. Stipules deciduous.

### Family 1. CUPULIFERÆ.

Epigynous. Style 1; stigmas several. Fruit a nut (akene), the base of which is surrounded by a cupula or a persistent involucre.

## 1. Quercus L.  Oak.

Monœcious. Aments slender. ♂ perigonium 4–8-parted; ♀, 6-dentate; style short; stigmas 3. Ovary 3-celled, 6-ovulate, surrounded by a scaly, budlike involucre, which in fruit enlarges into a cup.— ♃.

1. 2. LOBATA Née. ♂: aments pendulous; abortive ovules at the base of the developed seed; acorns maturing the first year; leaves deciduous; branchlets glabrous; nut conical. Common. Spring.

"White oak." "Roble." Large, stately tree.

2. Q. DOUGLASII Hook. & Arn. ♂: aments pendulous; abortive ovules at the base of the developed seed; acorns maturing the first year; leaves deciduous; branchlets pubescent; nut oblong.—Niles. Sonoma.

"Blue oak."

3. Q. DUMOSA Nutt. ♂: aments pendulous; abortive ovules at the base of the developed seed; acorns maturing the first year; leaves persistent, coriaceous; nut oval.—Tamalpais.

4. Q. CHRYSOLEPIS Liebm. ♂: aments pendulous; abortive ovules scattered over the surface of the developed seed; acorns biennial; leaves persistent, coriaceous; nut obtuse.— Marin County. Spring.

Tall tree, frequently called live oak.

5. Q. AGRIFOLIA Née. ♂: aments pendulous; abortive ovules apical; acorns maturing the first year; leaves persistent, coriaceous, not reticulate; nut acute.—Common. Spring.

Tall tree, the genuine "live oak" of the Californians.

6. Q. WISLIZENI DC. ♂: aments pendulous; abortive ovules apical; acorns biennial; leaves persistent, coriaceous, reticulate, entire to sinuately lobed; petioles short; nut acute. Niles. Spring.

Stately tree, frequently mistaken for "live oak."

7. Q. KELLOGGII Newb. ♂: aments pendulous; abortive ovules apical; acorns biennial; leaves deciduous, pinnatifid-lobed; petioles long; nut obtuse.—Marin County. Spring. "Black Oak."

8. Q. DENSIFLORA Hook. & Arn. Aments erect, androgynous.— Marin County. Santa Cruz mountains. Summer. "Chestnut." "Tanner's bark."

## 2. Castanopsis Spach. CHINQUAPIN.

Monœcious. Aments slender, panicled on the young shoots. ♂ perigonium 5-lobed; stamens 10. ♀ flowers 1–3, in a scaly involucre, sessile at the base of the ament; perigonium 6-lobed, 2-seriate; fruit 1–3 nuts entirely closed in a prickly, irregularly rupturing involucre.— ♄.

1. C. CHRYSOPHYLLA A. DC.—Tamalpais. Santa Cruz mountains. Summer.

In Northern California a tall tree, in our local flora, a shrub.

### 3. Corylus Tourn. HAZEL.

Monœcious. Aments slender. Scales (bracts) imbricated. ♂ flowers besides the scales 2-bracteolate. Stamens 8; anthers 1-celled. ♀ flowers immersed into a bud; uppermost scales containing flowers, the rest empty. Ovary 2-bracteolate, crowned by a minute perigonium, 2-celled; stigmas 2. Fruit a nut, surrounded by a large incised involucre, formed by the growth of the two bractlets.— ♄.

1. C. ROSTRATA Ait.—Common. Spring.

### Family 2. BETULACEÆ.

Hypogynous. ♂ ament: bracts shield-shaped, each enclosing 3 flowers with minute perigonium and four stamens. ♀ aments: bracts 3-lobed, each enclosing 3 naked flowers; ovary 2-celled; cells 1-ovulate; styles 0; stigmas 2. Fruit an akene (nutlet) affixed to the lignescent bract, and collected into a strobilaceous inflorescence.

### 1. Alnus Tourn. ALDER.

Fruit lignescent; nutlets attached to lignescent bracts on a lignescent axis.— ♄.

1. A. RUBRA Bongard. Nutlets w i n g e d. Contra Costa Mountains. Spring.

2. A. RHOMBIFOLIA Nutt. Nutlet not winged, but having a thickened margin.—Niles. Santa Cruz mountains. Spring.

### Family 3. MYRICACEÆ.

Flowers diclinous, amentaceous, each bract containing a single, naked, sessile flower. ♂ aments filiform; ♀ ovate. Ovary connate with some hypogynous scales, 1-celled, 1-ovulate; style very short; stigmas 2. Fruit a nutlet; drupaceous by the incrassate, hypogynous bracts.

### 1. Myrica L. Wax-Myrtle.

Only genus.— ♄.

1. M. CALIFORNICA Cham.—Common. Spring.

ORDER 2. URTICALES. Flowers diclinous. Stamens opposite to the lobes of the perigonium. Ovary superior. Fruit 1-seeded. Leaves stipulate.

### Family 1. PLATANEÆ.

Monœcious. Flowers naked, capitulate on globose receptacles. ♂ head: stamens ∞, irregularly mingled with scales. ♀ head: ovaries ∞, placed irregularly among scales; fruit a coriaceous nutlet. Leaves alternate, palmately lobed.

### 1. **Platanus** L.   Sycamore.

Only genus.— ♄.

1. P. racemosa Nutt.—Common.   Spring.

### Family 2. URTICACEÆ.

♂ perigonium 4-lobed; stamens 4; filaments inflexed in æstivation; anthers introrse, 2-celled; ovary rudimentary. ♀ perigonium 2 to 4-lobed; ovary 1-celled, 1-ovulate; fruit an akene, sometimes baccate by the persistent and succulent perigonium. Stipules generally persistent.

### 1. **Urtica** Tourn.   Nettle.

♂ perigonium regular and spreading. ♀ perigonium 2-parted, with sessile, penicillate stigma. Herbs covered with stinging hairs.

1. U. holosericea Nutt. Inflorescence unisexual; petioles short, stout.—♃. Common. Summer.

2. U. Lyallii Watson. Inflorescence unisexual; petioles long, slender.—♃. Marin County. Summer.

3. U. urens L. Inflorescence androgynous. ⊙. Cultivated lands. Spring.
Introduced from Europe.

### 2. **Hesperocnide** Torr.

♀ perigonium gamophyllous, urceolate, compressed, minutely dentate.—⊙.

1. H. TENELLA Torr.—San Francisco. Saucelito. Spring.

Small stinging herb.

### Family 3. POLYGONACEÆ.

Perigonium imbricate in æstivation. Stamens definite, inserted on the base of the perigonium. Ovary superior, originally compound of 2 or 3 carpidia; by abortion 1-celled, 1-ovulate; but showing the 2 or 3 styles of the original carpidia. Ovule basilar. Fruit indehiscent. Stems articulate. Leaves alternate, sheathing.

[Affinities of this family not yet settled. The articulate stem approaches it to the *Piperaceæ*; the basilar ovule to the *Chenopodiaceæ*.]

#### 1. Rumex L.   DOCK. SHEEP-SORREL.

Perigonium 6-sepalous, 2-seriate; 3 inner segments larger and more petaloid. Stamens 6, in pairs opposite the outer sepals. Styles 3 with penicillate stigmas. Fruit a 3-cornered akene inclosed in the 3 inner sepals, forming a spurious capsule. Flowers racemose.

1. R. SALICIFOLIUS Weinm. Calyx valves entire or denticulate, bearing large callosities; leaves linear to lanceolate, not undulate, attenuate into a short petiole.—♃. Common. Spring.

2. R. CRISPUS L. Calyx-valves entire, bearing callosities; leaves lanceolate, undulate, truncate at base.—♃. Common. Spring.

"Yellow dock." Native of Europe, now naturalized.

3. R. CONGLOMERATUS Murray. Calyx valves entire, bearing callosities; leaves lanceolate, the lower leaves cordate, slightly undulate. ♃. Common. Spring.

Also called "yellow dock," native of Europe, and at present naturalized.

4. R. OBTUSIFOLIUS L. Calyx valves with slender awn-like teeth and bearing callosities. ♃. Contra Costa. Spring.

7. R. ACETOSELLA L. Diœcious. Valves without callosities; leaves hastate.—♃. Common. Spring. "Sheep-sorrel."

### 2. Polygonum L. KNOT-GRASS.

Perigonium 5, sometimes 4-phyllous. Stamens 3-8, single or in pairs opposite to the sepals. Styles 3, sometimes 2; stigmas capitulate. Fruit a 3-cornered or lenticular akene, enclosed in the persistent perigonium.

1. P. PARONYCHIA Cham. & Schlecht. Flowers in leafy spikes; leaves lanceolate, with revolute margins; flowers pale, veined with green. ♃. San Francisco. Summer.

2. P. AVICULARE L. Flowers axillary, pale. —☉. Common. Spring. Summer.

3. P. CALIFORNICUM Meissn. Flowers spicate; bracts foliaceous, each bract containing

a single sessile flower; leaves linear; flowers rose-color.—☉. Niles. Spring.

4. P. NODOSUM Pers. Flowers spicate; bracts scarious, each bract protecting a fascicle; stamens 6; styles 2, included; leaves cuneate at base; flowers pale.—☉. Marine Hospital. Spring.

5. P. MUHLENBERGII Watson. Flowers spicate; bracts scarious, each bract protecting a a fascicle; stamens 5; style 2-cleft, exserted; leaves cordate at base; flowers rose-color.—♃. Marine Hospital. Spring.

6. P. CONVOLVULUS L. Stem twining; leaves hastate; flowers pale.—☉. Cultivated grounds. Summer.

Ballast weed, introduced from Europe.

### 3. Eriogonum Michx.

Involucre ∞-flowered, campanulate, slightly angulate, 6-dentate. Flowers ☿ pedicillate, exserted, seated on a receptacle with scarious bracts. Perigonium 6-parted, 2-seriate. Stamens 9, in pairs opposite to the external lobes; singly to the internal ones. Styles 3; stigmas capitate. Fruit a 3-angled, rarely a lenticular, akene. Peduncles 2–3-chotomous.

1. E. ANGULOSUM Benth. Involucre nerveless, hemispherical; flowers pale.—☉. Niles. Spring.

2. E. LATIFOLIUM Smith. Involucres 5–6-nerved, collected into a glomerule; peduncles stout, solid; flowers rose-color.—♃. Common. Summer.

3. E. NUDUM Dougl. Involucres 5–6-nerved, collected into a glomerule; peduncles slender, fistulose; flowers pale. — ♃. Coast Range. Summer.

4. E. TRUNCATUM Torr. & Gray. Involucre 5 or 6-nerved, solitary, forming a cyme; leaves rosulate, mostly radical; flowers rose-colored. ⊙. Niles. Spring.

5. E. VIRGATUM Benth. Involucre 5 or 6-nerved, solitary, forming a dichotomous panicle; leaves lanceolate; bracts shorter than the involucre; flowers pale.—⊙. Coast Range. Spring.

6. E. VIMINEUM Dougl. Involucre 5 or 6-nerved, solitary, forming a dichotomous panicle; leaves orbicular; flowers pale.—⊙. Common. Spring.

7. E. GRACILE Benth. Involucre 5 or 6-nerved, solitary, forming a dichotomous panicle; leaves lanceolate; bracts longer than the involucres; flowers pale. — ⊙. Santa Cruz mountains. Spring.

### 3. Chorizanthe R. Br.

Involucre 3-flowered, tubular, 3-angular, 6-dentate, mucronate or aristate. Flowers ☿, scarcely exserted. Perigonium 6-parted, 2-seriate. Stamens typically 9, in pairs opposite to the external lobes; singly to the internal ones; occasionally by abortion 6 or 3. Styles 3; stigmas capitate. Fruit a 3-angular akene. Peduncles dichotomous.

1. C. MEMBRANACEA Benth. Stem erect; involucre equally 6-cleft, 3-flowered, only one flower developed; leaves scattered.—⊙. Niles. Spring.

2. C. DOUGLASII Benth. Stem erect; involucre unequally cleft, 1-flowered; leaves verticillate.—⊙. Santa Cruz mountains. Spring.

3. C. PUNGENS Benth. Stem procumbent; involucre unequally dentate, teeth alternately smaller; leaves mostly opposite.—⊙. Presidio. Spring.

### 5 Lastarriæa Remy.

Involucre 0. Perigonium coriaceous, tubular, 6-dentate, the dentations uncinately aristate; stamens 3, inserted on the throat. Fruit a 3-angled akene.—⊙. Diffusely branched.

1. L. CHILENSIS Remy.—Common. Spring.

### 6. Pterostegia Fisch. & Mey.

Involucre a single 2-lobed bract, subtending a single ☿ flower. Perigonium 6-parted, persistent. Stamens 6 (rarely fewer). Fruit a 3-angular akene, loosely enveloped by the enlarged involucre. Prostrate, diffusely dichotomous; leaves opposite.—☉.

1. P. DRYMARIOIDES Fisch & Mey. Common. Spring.

ORDER 3. **PIPERALES.** Flowers sessile, in dense spikes or racemes, bracteate, without perigonium. Stem articulate.

### Family 1. SAURUREÆ.

Ovary central, compound of several follicular carpidia; stigmas several.

1. **Anemopsis** Nutt.   YERBA MANSA.

Flowers in a dense, conical spadix, with a several-leaved, persistent, colored involucre; each flower subtended by a colored bract. Stamens adnate to the base of the ovary. Ovary immersed into a rachis, 1-celled, with parietal placentæ. Fruit a capsule, dehiscent from the apex. Leaves mostly radical.—♃.

1. A. CALIFORNICA Hook. Bracts and involucre white.—San Pablo marshes. Summer.

Herb and root-stock aromatic.

### Family 2. CERATOPHYLLEÆ.

Monœcious. Involucre 12-cleft; lobes linear, truncate. ♂ flowers: anthers ∞, sessile. ♀ flowers: ovary 1-celled, 1-ovulate; style terminal; fruit a nutlet, with persistent style and involucre. ♃. Submersed aquatics; stem articulate; leaves verticillate, sessile, dichotomous.

Affinities doubtful. Further investigations even may remove them altogether from the *Angiosperms* and place them near the *Gnetaceæ*, together with the *Podostemoneæ*.

### 1. Ceratophyllum L.

Only genus of the family.

1. C. DEMERSUM L. Common aquatic.

## Division 2. MONOCOTYLEDONES.

### Series 1. COROLLIFLORÆ.

Perigonium 2×3 parted. Ovary 3 carpidia, syncarpous in different degrees.

#### Order 1. GYNANDRÆ.

#### Family 1. ORCHIDEÆ.

Placentæ parietal.

##### 1. Epipactis Haller.

Anther 1, persistent, not connate with the column; pollinia 2, attached to the common gland. Perigonium spreading; labellum geniculate, inferior part concave. Ovary contorted only at base; column short, terete.—♃ Caulescent, flowers in a loose, few-flowered raceme.

1. E. GIGANTEA Dougl. Sepals brownish-green, labellum white, dotted.—San Francisco [extinct]. Camp Taylor. Summer.

## 2. Spiranthes Richard.  LADIES' TRESSES.

Anther 1, persistent, not connate with the column; pollinia 2, attached to the common gland. Perigonium oblique ; labellum enclosed, canaliculate, embracing with its base the short column.—♃. Flowers forming a spiral spike.

1. S. ROMANZOFFIANA Cham. Callosities of the lip smooth, obscure; flowers greenish-white.—Presidio. Marin County. Autumn.

2. S. PORRIFOLIA Lindl. Callosities of the lip prominent, pointing downward ; flowers greenish-white.—Marin County. Autumn.

### 3. Habenaria Willd.

Anther 1, persistent, entirely connate with the column; pollinia 2, divergent at the lower end. Perigonium ringent, galeate ; lip elongate, spreading, calcarate. Column short.— ♃. Caulescent.

1. H. ELEGANS Bolander. Sepals 1-nerved; flowers green. — Oakland hills. San Mateo. San Francisco [extinct]. Summer.

2. H. LEUCOSTACHYS Watson. Sepals 3-nerved ; flowers white. — Saucelito. San Francisco [extinct]. Summer.

### 4. Corallorhiza Haller. CORAL-ROOT.

Anther 1, terminal, caducous, not adnate to to the column; pollinia 4, globose. Perigonium ringent. Lip adnate to the column, serrate at base, 3-lobed; lateral lobes very small. Column semi-terete.—♃. Aphyllous and without chlorophyll; flowers in spikes.

1. C. MULTIFLORA Nutt. Flowers calcarate; sepals greenish; lip white, dotted with red.—Crystal Springs. San Mateo. San Francisco [extinct]. Summer.

2. C. BIGELOVII Watson. Spur 0, sepals purple.—Tamalpais. Summer.

### 5. Cypripedium L. LADY'S SLIPPER.

Anthers 2. Lip inflated.—♃. Flowers few or solitary.

1. C. MONTANUM Dougl. Sepals brownish; lip pale, veined with purple.—Santa Cruz Mountains. Summer.

ORDER 2. EPIGYNÆ. Stamens free, 3 or 6. Ovary inferior.

### Family 1. IRIDACEÆ.

Stamens 3, opposite to the external segments of the perigonium; anthers extrorse.

### 1. Sisyrinchium L. BLUE-EYED GRASS.

Divisions of the perigonium equal. Stamens monadelphous. Style short; stigmas 3, filiform, involute, alternate with the stamens. Fruit a capsule, obovate.

1. S. BELLUM Watson. Filaments united to the top; flowers blue with yellow center.—♃. Common. Summer.

2. S. CALIFORNICUS Ait. Filaments united only at the base; flowers yellow.—♃. Colma. Saucelito. Summer.

## 2. Iris L.

Perigonium tubular at the base; segments equal, but differing in shape; the 3 external reflexed; the 3 internal erect. Stigma 3-parted, petaloid.—♃.

1. I. MACROSIPHON Torr. Tube of the perigonium elongated, cylindrical; stem leafy; capsule ovoid, acute at each end; flowers variable in color, but most frequently lilac-purple. Marin County. Spring.

2. I. DOUGLASIANA Herbert. Tube of the perigonium elongated, cylindrical; stem leafy; capsule oblong, 3-gonal; flowers pale lilac, external sepals with white center, lined with purple.—Common. Spring.

3. I. LONGIPETALA Herbert. Tube of the perigonium short and funnel-shaped; stems almost naked; flowers lilac, external sepals white, veined with violet; midrib yellow. Common. Spring.

ORDER 3. CORONARIÆ. Perigonium 6-parted in different degrees. Stamens 3 or 6. Ovary superior.

## Family 1. SMILACEÆ.

Anthers not extrorse. Ovary 3-celled, with central placentæ. Fruit a berry. Testa of the seed not crustaceous.

### 1. Smilacina Desf.  FALSE SOLOMON'S SEAL.

Divisions of the perigonium equal. Stamens inserted at the base of the segments. Cells of the ovary 2-ovulate; cells of the ripe fruit 1-seeded.—♃. Leaves sessile and amplexicaul.

1. S. AMPLEXICAULIS Nutt. Flowers greenish-white, panicled; stamens exceeding the perigonium; berry red.—Common. Spring.

2. S. SESSILIFOLIA Nutt. Flowers in a simple raceme, white; stamens shorter than the perigonium; berry dark.—Common. Spring.

### 2. Maianthemum Mœnch.

Perigonium 4-parted. Stamens 4.—♃. Leaves petiolate.

1. M. BIFOLIUM DC. Flowers white; berry red.—Saucelito. Spring.

### 3. Prosartes Don.

Perigonium 6-sepalous, campanulate; segments saccate at base. Stamens 6, attached at the base of the sepals, and deciduous with them. Anthers adnate, dehiscing laterally. Ovary 3-celled; cells 2-ovulate; style deciduous. Cells of the berry 1-2-seeded. — ♃. Leaves alternate, sessile, amplexicaul.

1. P. MENZIESII Don. Perigonium gibbous at base, its sepals almost erect; flowers green; berry red.—Lagunitas Creek. Spring.

2. P. HOOKERI Torr. Perigonium narrow at base, its sepals spreading above; flowers green; berry red.—Common. Spring.

### 3. Trillium L.  WAKE-ROBIN.

Perigonium 6-sepalous; 3 external sepals herbaceous, persistent; 3 internal petaloid, marcescent. Stamens 6; anthers linear, lateral; connective broad. Ovary 3-celled, $\infty$-ovulate; styles 3. Berry $\infty$-seeded.—♃. Leaves 3, verticillate; nervation reticulate.

1. T. SESSILE L. Flowers sessile; corolla variable; berry red.—Common. Spring.

Rhizome, and probably also berry, an acrid poison.

2. T. OVATUM Pursh. Flowers pedunculate, white, gradually turning red.—Tamalpais. Camp Taylor. Spring. Poisonous.

### Family 2. LILIACEÆ.

Anthers not extrorse. Ovary 3-celled. Testa of the seed crustaceous.

### 1. Clintonia Raf.

Perigonium 6-sepalous, campanulate, deciduous. Stamens 6, inserted on the base of the sepals; anthers versatile. Style slender, deciduous. Fruit a berry.—♃. Flowers on a scape.

1. C. ANDREWSIANA Torr. Flowers scarlet; berry blue.—Los Gatos. Lagunitas Creek. Summer.

### 2. Chlorogalum Kunth.   SOAP-ROOT.

Perigonium 6-sepalous; sepals linear, persistent, marcescent. Stamens 6, adnate to the base of the sepals; anthers versatile. Cells of the ovary 2-ovulate; style filiform, deciduous. Fruit a membranaceous, 3-lobed, loculicidal capsule. — ♃. Bulbous; inflorescence racemose, paniculate.

1. C. POMERIDIANUM Kunth. Flowers white. Common. Summer.

In former times used in washing linen. Old settlers consider it beneficial to the growth of the hair, like the bark of the Chilean Rosaceous tree " *Quillaja*."

### 3. Allium L.   ONION.

Perigonium persistent. Stamens 6, adnate to the base of the segments; filaments naked with dilated base; anthers versatile. Ovary 3-lobed; cells 2-ovulate; style filiform, persistent; stigma simple. Fruit a loculicidal capsule.—♃; bulbous; scapigerous; inflorescence umbellate.

1. A. UNIFOLIUM Kellogg. Capsule not crested; leaves linear, flat; flowers rose-color. Tamalpais. Berkeley. Summer.

2. A. ATTENUIFOLIUM Kellogg. Capsule 6-crested; leaves convolute, filiform; perigonium at last thin and lax; flowers pale.—Colma. Marin County. Summer.

3. A. SERRATUM Watson. Capsule 6-crested, crests narrow, central; perigonium fleshy, rigid; flowers rose-color. — Colma. Lake Chabot. Summer.

4. A. LACUNOSUM Watson. Capsule bearing toward the summit 6 obtuse ridges. Perigonium fleshy, rigid; flowers rose-color.—Alma. Summer.

### 4. Cyanotris Raf.
(*Camassia* Lindl.)   CAMASS.

Perigonium 6-sepalous, persistent; 5 superior sepals ascending; inferior one deflexed. Filaments filiform, ascending; anthers versatile. Ovary ovate, 3-celled; cells $\infty$-ovulate; style filiform, declinate; stigma slightly 3-cleft. Fruit a loculicidal, 3-angulate capsule; cells $\infty$-seeded.—♃. Bulbous; scapigerous; flowers in a simple raceme.

1. C. ESCULENTA Lindl. Flowers of different shades of blue.—Marin County. Summer.

The bulb has a taste like garlic, and is eaten by the Indians.

### 5. Brodiæa Smith.

Perigonium funnel-shaped, angulate, persistent. Stamens 3, alternating with as many staminodia. Hypogynous disc 3-lobed. Ovules ∞. Style persistent; stigma 3-lobed. Capsule loculicidal.—♃. Scapigerous; umbellate.

1. B. GRANDIFLORA Smith. Leaves about equaling the scape; cells of the ovary 6–8-seeded; flowers purple to rose-color.—Common. Summer.

2. B. MINOR Watson. Cells of the ovary 3-seeded; flowers purple.—Marin County. Summer.

3. B. TERRESTRIS Kellogg. Leaves considerably longer than the short scape; cells of the ovary 6–8-seeded; flowers purple.—Crystal Springs. Summer.

4. B. CONGESTA Smith. Flowers subcapitate; segments of perigonium twice longer than the tube; flowers purple.—Common. Summer.

### 6. Triteleia Hook.

Perigonium salver-shaped, persistent. Fertile stamens 6; filaments short, 3 inserted on the throat, and 3 half way down the tube. Ovary stipitate; ovules ∞. Capsule loculicidal.—♃. Scapigerous umbellate; capitate.

1. T. CAPITATA Benth. Stamens in 2 rows; flowers subcapitate; segments of perigonium

little longer than the tube; flower blue, varying from purple to white.—Common. Summer.

2. T. LAXA Watson. Stamens in 2 rows; flowers umbellate; tube of perigonium exceeding the lobes; flowers blue.—Common. Summer.

3. T. PEDUNCULARIS Watson. Stamens in 2 rows; flowers umbellate; lobes of perigonium exceeding the tube; flowers purple, sometimes white.—Tiburon. Summer.

4. T. IXIOIDES Watson. Stamens almost in a simple row, inner filaments exceeding the 3 outer ones; flowers yellow, marked with purple.—San Rafael. Summer.

5. A. LACTEA Watson. Stamens almost in a simple row; all six filaments alike; flowers pale with green nerve.—Common. Summer.

### 7. Muilla Watson.

Perigonium subrotate, persistent. Stamens 6, inserted near the base; anthers versatile. Ovary sessile, $\infty$-ovulate; style persistent, clavate, at length splitting. Capsule globose, loculicidal.—♃. Scapigerous; umbellate.

1. M. MARITIMA Watson. Sepals white with 2 green nerves.—Common. Summer.

### 8. Lilium L. LILY.

Perigonium 6-sepalous; sepals equal, spreading with a nectariferous groove near the base, deciduous. Ovary ∞-ovulate; style undivided; stigma 3-angular.—♃. Bulbous. **Pedicels bractless.**

1. L. PARDALINUM Kellogg, Flowers orange-color, with purple dots.—Marin County. Contra Costa. Summer.

### 9. Fritillaria L.

Perigonium deciduous; 6-sepalous, campanulate, with a smooth nectariferous pit near the base of each sepal. Ovary ∞-ovulate; style 3-parted.—♃. Bulbous. Flowers bracteate.

1. F. LILIACEA Lindl. Leaves in a single whorl near the base of the stem; capsule stipitate, with obtuse angles; flowers greenish-white.—South San Francisco. Mission Hills. Lake Chabot. Summer.

2. F. LANCEOLATA Pursh. Leaves in 1 to 3 whorls on the upper part of the stem; flowers dark purple, mottled with yellow.—Common. Summer.

Bulb of this species probably poisonous, like that of its European congener *F. Meleagris.*

### 10. Calochortus Pursh.    MARIPOSA.

Perigonium 6-sepalous, deciduous; 3 external sepals smaller and lanceolate; 3 internal, broad, with a conspicuous nectariferous gland near the base. Ovary ∞-ovulate; style 0; stigmas 3, reflexed, persistent.—♃. Bulbous.

1. C. ALBUS Dougl. Flowers subglobose, nodding; glandular pit of sepals shallow; with 4 transverse scales; flowers white.—San Mateo. Marin County. Summer.

2. PULCHELLUS Dougl. Flowers subglobose, nodding; glandular pit of sepals deep and covered by stiff appressed hairs; flowers yellow. Napa. Summer.

3. C. LILACINUS Kellogg. Flowers pale-lilac; erect; pedicels slender, recurved in fruit. Oakland Resorvoir. Lagunitas. Summer.

4. C. LUTEUS Dougl. Flowers and capsules erect on stout pedicels; glands of the sepals rounded and about as broad as the claw; color of inner sepals variable (yellow, white or lilac) marked with purple.—Common. Summer.

5. C. VENUSTUS Benth. Flowers and capsules erect on stout pedicels; gland of the inner sepals much narrower than the claw; color of inner sepals white or lilac, with a reddish spot near the top, and a purple spot bordered with yellow.—Contra Costa Range. Summer.

### Family 3. MELANTHACEÆ.

Anthers extrorse. Ovary 3-celled; styles 3. Capsule septicidal. Testa of the seeds not crustaceous.

### 1. Scoliopus Torr.

Perigonium 6-sepalous, spreading, deciduous; 3 external sepals lanceolate; 3 internal linear. Stamens 3, inserted at the base of the external sepals; anthers 2-celled, attached above the base. Ovary 3-quetrous, 1-celled, with 3 parietal placentæ; ovules 2-seriate on each; styles 3, linear, persistent, recurved, canaliculate. Capsule irregularly dehiscent.—♃. Leaves 2. Peduncles long, 1-flowered, arranged in an umbel.

1. S. Bigelovii Torr. Outer sepals dark purple, inner sepals pale, lined with purple. Saucelito. Tamalpais. Spring.

### 2. Zygadenus Michx.   Zygadene.

Perigonium 6-sepalous; sepals unguiculate, 2-glandular at the base, persistent; stamens 6, inserted on the claw, anthers reniform. Ovary 3-celled, ∞-ovulate, styles 3, divergent; capsule septicidal.—♃. Bulbous. Leaves lanceolate. Inflorescence racemose.

1. Z. Fremontii Torr. Outer sepals unguiculate; flowers greenish-white.—Marin County. Crystal Springs. Spring.

Has the reputation of being poisonous.

2. Z. venenosus Watson. All sepals unguiculate; flowers greenish-white.—Common. Spring.

Has the reputation of being poisonous, but is certainly not so to hogs, who devour its bulbs greedily and without any injury to themselves.

### 3. Xerophyllum Rich.

Perigonium 6-sepalous; sepals sessile, persistent; 3 internal ones smaller. Stamens inserted on the base of sepals; anthers laterally dehiscent. Ovary 3-celled; cells few-ovulate; styles 3, recurved, persistent. Capsule septicidal.—♃. Rhizomatous. Leaves linear, elongate. Inflorescence racemose.

1. X. TENAX Nutt. Flowers white.--Tamalpais. Wright's Station. Spring.

This plant is remarkable by its flowering plentifully at certain years and then disappearing altogether for a long period. This circumstance is probably owing to the fact that the plant requires a fixed age before flowering and then dies after having ripened its seeds. We find the same peculiarity in many species of Bamboo, in some palms, viz.: *Corypha*, and in *Agave*, all of which plants are annuals in a botanical sense, but require a number of years to complete their biological cycle.

### Family 4. JUNCACEÆ.

Perigonium glumaceous. Anthers basifixed.

## 1. Juncus L.   Bog-Rush.

Perigonium 6-sepalous; 3 external sepals carinate. Ovary 3-celled; cells $\infty$-ovulate. Capsule 3-valved, with central placentæ, $\infty$-seeded. Mostly aquatics.

1. J. ROBUSTUS Watson. Stem naked. Inflorescence lateral; flowers of the panicle branches clustered; leaves terete and pungent. ♃. Santa Clara. Summer.

2. J. LESEURII Bolander. Stem naked, terete; inflorescence lateral; flowers of the compound panicle branches solitary; leaves 0; capsule oblong, acute, but not rostrate.—♃. Presidio. Summer.

3. J. BALTICUS Dethard. Stem naked, terete; inflorescence lateral; flowers of the compound panicle branches solitary; leaves 0; capsule oblong, rostrate.—Common. Summer.

4. J. COMPRESSUS HBK. Stem naked, somewhat flattened; inflorescence lateral; flowers of the loosely few-flowered panicle solitary; leaves terete, sometimes 0; capsule oblong. ♃. Santa Clara. Summer.

5. J. BREWERI Engelmann. Stem naked, somewhat flattened; inflorescence lateral; flowers of the small, but dense panicle, solitary on their branches; leaves 0; capsule oblong.—♃. Santa Cruz. Summer.

6. J. EFFUSUS L. Stem naked; inflorescence lateral; flowers of the panicle solitary on their branches, triandrous; leaves 0; capsule clavate obtuse.—♃. Presidio. Summer.

7. J. PATENS Meyer. Stem naked; inflorescence lateral; flowers on panicle branches solitary, hexandrous; leaves 0; capsule subglobose. ♃. Common. Summer.

8. J. BUFONIUS L. Stem leafy, branched; inflorescence terminal; flowers remote, hexandrous.—⊙. Common. Summer.

9. J. KELLOGGI Engelmann. Stem leafy, branched; inflorescence terminal; peduncles 1 or 2-flowered; flowers triandrous.—⊙. San Francisco. Spring.

10. J. TENUIS Willd. Stem naked, simple; inflorescence terminal, a diffuse cyme of solitary flowers.—♃. Lake Chabot. Summer.

11. J. FALCATUS Meyer. Stem leafy; leaves sheathing, flat.—♃. Lone Mountain. San Francisco. Summer.

12. J. XIPHIOIDES Meyer. Stems leafy, ancipital; leaves equitant, laterally compressed; flowers capitate; heads paniculate; anthers exceeding the filaments.—♃. Coast Range. Summer.

13. J. PHÆOCEPHALUS Engelm. Stem leafy, ancipital; leaves equitant, laterally compressed; flowers capitate; heads paniculate; anthers ex-

ceeding the filaments.—♃. Coast Range. Crystal Springs. Summer.

### 2. **Luzula** DC.   WOOD-RUSH.

Perigonium 6-sepalous, all the segments flat. Ovary 1-celled, 3-ovulate. Capsule 3-valved, 3-seeded.—♃.

1. L. COMOSA Meyer. Common. Spring.

SERIES 2. MICRANTHÆ. Flowers inconspicuous. Inflorescence ∞-flowered.

ORDER 1. GLUMACEÆ. Flowers in the axils of bracts, and arranged in spikelets. Spathes 0. Perigonium depauperate. Ovary 1-celled, 1-seeded. Fruit a caryopsis, with endosperm.

#### Family 1. **GRAMINEÆ**.

Stem articulate. Leaves distichous, alternate, sheathing, with ligule (stipule) at base. Flowers protected by an anterior and a posterior bractlet (palet). Spikelet generally protected by one or two bracts (glumes). Stamens usually 3. Pistils 2.

### 1. **Panicum** L.   PANIC-GRASS.

Spikelets 2-flowered; upper one ☿, lower reduced to a single palet. Glumes unequal, the lower smaller. Leaves flattened.

1. P. SANGUINALE L. Spikelets in pairs; one sessile, the other pedicillate; crowded on

one side of the simple flattened branches, which are digitately clustered at the top of the culm.—☉. Common. Summer.

Probably introduced from Europe.

2. P. AGROSTOIDES Spreng. Spikelets disposed in panicles; panicle elongated, racemose. ♃. San Jose. Summer.

3. P. CAPILLARE L. Panicle diffuse and spreading; spikelets scattered, disposed in panicles and pointed.—☉. San Francisco. Summer.

4. P. DICHOTOMUM L. Panicle diffuse and spreading. Spikelets scattered, disposed in panicles and obtuse.—☉. Marin County. Summer.

5. P. CRUSGALLI L. Spikelets crowded on the secund, spikelike branches of the panicle. ☉. Common. Summer.

Introduced from Europe.

## 2. Phleum L.  TIMOTHY.

Spikelets 1-flowered, ☿. Glumes 2; aristate; the upper palet sometimes bearing at its base the rudiment of a second flower. Leaves flattened. Panicles spikelike.

1. P. PRATENSE L. Awn shorter than its glume.—♃. Common. Summer. Native of Europe.

2. P. ALPINUM L. Awn about as long as its glume.—♃. San Francisco. Summer.

### 3. Alopecurus L.    Foxtail.

Spikelet 1-flowered, ☿. Glumes 2, navicular, connate at their base; lower palet carinate; upper wanting.—♃. Leaves flattened. Panicles spikelike.

1. A. pratensis L. Culm erect.—Niles. Summer.

2. A. geniculatus L. Culm procumbent, ascending from the lower nodes.—Marin County. Summer.

### 4. Beckmannia Host.    Caterpillar Grass.

Spikelets 1 or 2-flowered. Flowers ☿, sessile. Glumes 2, navicular; lower palet ovate, 3-nerved, embracing the upper 2-cleft, 2-nerved, one.—♃. Leaves flattened. Spike sessile; spikelets alternate, sessile, 1-sided, 2-seriate.

1. B. erucæformis Host.—Marin County. Summer.

### 5. Phalaris L.    Canary Grass.

Spikelets 3-flowered, upper ☿, 2 lower neutral, depauperate. Glumes 2, carinate. Palets 2, navicular; the lower larger and embracing the upper. Spikelets pedicillate. Leaves flattened.

1. P. Canariensis L. Panicle dense and spikelike; glumes broad with a broad keel.—☉. Common near houses. Spring.

"Canary grass." Native of the Mediterranean region; used as bird's food and probably escaped with the rubbish of birds' cages.

2. P. INTERMEDIA Bosq. Panicle dense and spike-like; glumes pointed with a broad keel. ⊙. Common. Spring.

3. P. AMETHYSTINA Trin. Panicle dense and spike-like; glumes with a narrow keel.—♃. Contra Costa hills. Summer.

4. P. ARUNDINACÆ L. Panicle branched; glumes not all carinate, pointed, 3-nerved.—♃. Niles. Summer.

### 6. Hierochloa Gmelin. HOLY GRASS.

Spikelets 3-flowered; flowers sessile. Glumes equal, carinate. Upper flower ☿; lower palet carinate; upper 1-nerved. 2 lower flowers ♂; lower palet carinate, upper 2-carinate.—♃. Spikelets pedicillate. Leaves flattened.

1. H. MACROPHYLLA Thurb.—Tamalpais. Summer.

Fragrant grass, perhaps containing cumarine like our *Galium triflorum* and the European *Asperula odorata* (Waldmeister).

### 7. Anthoxanthum L. SWEET VERNAL-GRASS.

Spikelets 3-flowered. Glumes 2, carinate; lower glume shorter, 1-nerved; upper 3-nerved. Upper flower ☿, with 2 navicular palets without awns; lower palet wrapt round the upper.

Stamens 2. Lower 2 flowers neutral, with a single, aristate, canaliculate palet.—♃. Leaves flattened. Panicle contracted.

1. A. ODORATUM L.—Marin County. Summer.

Fragrant grass, probably introduced from Europe.

### 8. Polypogon Desf. BEARD GRASS.

Spikelets 1-flowered. Glumes 2, carinate, aristate, much longer than the flower. Palets 2, upper 2-carinate, lower truncate at the apex. Leaves flattened.

1. P. MONSPELIENSIS Desf. Glumes notched, their awns 2 or 3 times their own length.—☉. Coast Range. Spring.

2. P. LITTORALIS Smith. Glumes tapering into an awn about their own length.—♃. San Francisco. Summer.

### 9. Agrostis L. BENT-GRASS.

Spikelets 1-flowered. Glumes 2, carinate, awnless, larger than the flowers. Palets 2; lower sometimes aristate; upper 2-carinate, sometimes wanting. Tufted. Leaves sometimes involute. Panicles with verticillate branches.

1. A. ALBA L (*vulgaris* With). Glumes nearly equal, upper palet nearly half the length of the lower.—♃. Common, although not indigenous. Summer.

Native of Europe.

2. A. VERTICILLATA Vill. Glumes nearly equal; upper palet nearly as long as the lower. ♃. San Francisco. Summer.

3. A. EXARATA Trin. Glumes nearly equal; upper palet minute, about the length of the ovary; lower palet 5-nerved and marked on the back by a longitudinal furrow.—♃. San Francisco. Summer.

4. A. SCABRA Willd. Glumes unequal, the lower longer, acute and scabrous; upper palet very minute, sometimes 0.—⊙. San Franciscco. Spring.

### 10. Gastridium Pal.

Spikelets 1-flowered. Glumes 2, ventricose at the base, much larger than the flower and closed. Palets 2; the inferior sometimes aristate, embracing the superior 2-carinate one.—⊙. Leaves flattened. Panicles contracted, spikelike.

1. G. AUSTRALE Beauv.—San Francisco. Spring.

### 11. Stipa L. FEATHER-GRASS.

Spikelets 1-flowered, flowers stipitate. Glumes 2, membranaceous, larger than the flower. Palets 2, involute, superior shorter, 2-nerved; inferior aristate, with a simple twisted awn. Ovary stipitate; caryopsis terete, and closely wrapt in the palets.—♃. Spikelets pedicillate, paniculate.

1. S. SETIGERA Presl. Panicle open, with spreading rays; lower palet tuberculate, only the nerves hairy.—Common. Summer.

2. S. EMINENS Cas. Panicle open, with spreading rays; lower palet hairy all over.—Coast Ranges. Summer.

3. S. VIRIDULA Trin. Panicle narrow, with short erect rays.—Coast Range. Summer.

All the species of Stipa are by their long awns more or less injurious to the wool of sheep, which they make impure and intractable. Some species even endanger the life of the animals, as their awns are liable to work their way from the wool through the skin into the lungs.

### 12. Deyeuxia Clar. (*Calamagrostis* Adans).

Spikelets 1-flowered, flowers sessile, with a bearded base and the rudiment of a superior flower, reduced to a plumose pedicel. Glumes 2, canaliculate, awnless. Palets 2; inferior aristate, superior 2-carinate.—♃. Leaves flattened. Panicles branched.

1. D. ALEUTICA Trin.—San Francisco. Summer.

### 13. Spartina Schreb. MARSH GRASS.

Spikelets 1-sided, sessile, 1-flowered; flower naked, sessile. Glumes 2, carinate, awnless; the upper embracing the lower, which is much smaller. Palets 2, awnless; the lower compressed; the upper larger navicular. Ovary sessile.—♃. with creeping rhizome, rigid, and with involute leaves.

1. S. STRICTA Roth.—Salt marshes. San Francisco. Summer.

### 14. Cynodon Rich.  BERMUDA GRASS.

Spikelets 1-sided, sessile, 1-flowered, often with rudiment of a superior flower. Glumes 2, carinate, awnless, the upper embracing the lower. Palets 2; the lower carinate, acute. Ovary sessile.—♃. Rhizome creeping. Leaves flattened. Spikes in our species digitate.

1. C. DACTYLON Pers.—San Jose. Summer.

### 15. Danthonia DC.

Spikelets 2-∞-flowered; rachis hairy. Flowers distichous, the uppermost depauperate. Glumes 2, awnless, somewhat longer than the flower. Palets 2; upper 2-carinate; lower concave, ∞-nerved, 2-cleft, aristate. Ovary stipitate; caryopsis compressed, free.—♃. Turfy. Leaves flattened. Spikelets pedicellate, paniculate.

1. D. CALIFORNICA Bolander.—Contra Costa hills. Summer.

### 16. Avena L.  OAT.

Spikelets 3-flowered, the uppermost depauperate. Glumes 2, awnless. Palets 2; lower 2-cleft, aristate; upper 2-carinate, awnless.

Ovary sessile, hirsute at the apex; caryopsis terete, adherent to the upper palet.—☉.

1. A. FATUA L.—Common. Summer.

### 17. Trisetum Kunth.

Spikelets 2-4-flowered. the uppermost depauperate. Glumes 2, carinate, awnless, shorter than the flower. Palets 2; lower 2-dentate, aristate. Ovary sessile; caryopsis compressed, free. Leaves flattened.

1. T. CANESCENS Buckl. Spikelets 2 or 3-flowered, narrow; lower glume narrow, considerably shorter than the broad ovate upper one.—☉. Common. Spring.

2. T. BARBATUM Steud. Spikelets 3-5-flowered, large, much flattened; both glumes narrow, the lower but little shorter than the upper one.—☉. Common. Spring.

### 18. Aira L.   HAIR-GRASS.

Spikelets 2-flowered, often with the rudiment of a superior one. Flowers sessile. Glumes 2, carinate, awnless, larger than the flower. Palets 2; the lower 2-cleft, dorsally aristate. Ovary sessile; caryopsis free.

1. A. CÆSPITOSA L. Glumes not longer than the florets; spikelets much compressed.—♃. San Francisco. Summer.

2. A. HOLCIFORMIS Steud. Glumes not longer than the florets; spikelets nearly terete.—♃. Common. Summer.

3. ELONGATA Hook. Glumes longer than the florets; panicle long and narrow, its rays unequal, distant, mostly appressed.—♃. San Francisco. Summer.

4. A. DANTHONOIDES Trin. Glumes longer than the florets; panicle loose and open, the lowermost rays in threes, the upper in pairs or solitary.—♃. Oakland. Summer.

### 19. Arrhenatherum Beauv.

Spikelets 3-flowered : lower ♂ ; middle ☿ ; upper neutral. Glumes 2, concave, awnless; the upper larger than the lower. Palets of ♂ flower 2; lower dorsally aristate near the base; upper 2-carinate, awnless. Palets of ☿ flower 2; lower dorsally aristate near the apex, upper adnate to the caryopsis.—♃. Leaves flattened. Panicles with verticillate branches. Spikelets pedicillate.

1. A. AVENACEUM Beauv.—Common. Summer.

European species, probably introduced as a fodder grass.

### 20. Holcus L.

Spikelets 2-flowered ; flowers pedicillate ; lower ☿ ; upper ♂. Glumes 2, navicular. Palets of the ☿ flower 2; lower navicular, awnless; upper 2-carinate. Ovary sessile, pyriform; caryopsis free. Palets of the ♂ flower 2; upper 2-carinate ; lower aristate near the apex. Leaves flattened. Panicles branched. Spikelets pedicellate.

1. H. LANATUS L.—♃. Common. Summer.

### 21. Phragmites Trin. REED.

Spikelets 3-6-flowered; flowers distichous, somewhat remote; the lower one ♂, the rest ☿. Glumes 2, carinate, acute; the upper larger. Palets 2; lower elongate, subulate, upper 2-carinate. Ovary sessile; caryopsis free.—♃. Aquatic. Leaves broad, flattened. Panicles diffused.

1. P. COMMUNIS Trin.— Common on river banks. Summer.

### 22. Dactylis L. ORCHARD GRASS.

Spikelets 1-sided, 2-7-flowered. Glumes 2, carinate, mucronate aristate. Palets 2; upper 2-carinate, lower 5-nerved, mucronate aristate; carina ciliate. Ovary sessile. Caryopsis free. ♃. Leaves carinate. Panicles glomerate, 1-sided.

1. D. GLOMERATA L.
Introduced from Europe as a fodder-grass.

### 23. Kœleria Pers.

Spikelets 2-7-flowered. Flowers distichous. Glumes 2, carinate, awnless, unequal. Palets 2; lower enlarged, upper 2-carinate, 2-cleft. Caryopsis terete, free. Leaves flattened. Panicles contracted, spikelike. Spikelets pedicellate.

1. K. CRISTATA Pers.—♃. San Francisco. Summer.

### 24. Melica L. MELIC-GRASS.

Spikelets 3-5-flowered, 2 inferior flowers ☿, the rest depauperate. Glumes 2, concave, awnless, unequal. Palets 2, sessile, smooth. Caryopsis terete, free. Leaves flattened. Spikelets pedicellate.

1. M. IMPERFECTA Trin. G l u m e s nearly equaling the florets; lower palet scarious margined, 7-nerved; spikelets one ☿ floret and one or two sterile.—♃. Cemetery. Summer.

2. M. BULBOSA Geyer. G l u m e s nearly equaling the florets; lower palet scarious-margined, 7-nerved; spikelets contain 2 or 3 ☿ florets.—♃. Coast Range Summer.

3. M. BROMOIDES Gray. Glumes distinctly shorter than the lower floret; lower palet acute. ♃. Marin County. Summer.

4. M. HARFORDI Bolander. Glumes distinctly shorter than the lower floret; lower palet truncate.—♃. Nicasio. Summer.

### 25. Brizopyrum Link. (*Distichlis* Raf.).

Diœcious. Spikelets compressed, ∞-flowered. Glumes 2, narrow, carinate. Palets 2: lower coriaceous, ∞-nerved, not carinate; upper carinate with involute margins. Ovary stipitate; caryopsis obovate, free.—♃. Creeping. Leaves rigid, distichous. spreading. Panicle spicate-racemose.

1. B. MARITIMUM Raf.—San Francisco. Salt marshes. Summer.

### 26. Lophochlæna Nees.

Spikelets ∞-flowered. Flowers ☿. Rachis articulate, deciduous. Glumes 2, shorter than the flowers, upper 3-nerved, lower smaller, 1-nerved. Palets 2: lower chartaceous, ∞-nerved, membranaceous at the apex, 2-lobed, aristate from the cleft; upper complicate, chartaceous, margin and apex membranaceous, 2-nerved, 2-carinate, margin dentate, apex emarginate. Caryopsis compressed, 2-horned. Leaves short. Panicle secund (one-sided), simply racemose.

1. L. CALIFORNICA Nees.—☉. Contra Costa. Spring.

### 27. Glyceria R. Br.

Spikelets ∞-flowered. Glumes 2, obtuse, the lower shorter. Palets 2: upper 2-carinate, lower concave, ovate, rotund, 7-nerved. Caryopsis free. Aquatics. Leaves flattened. Branches of the panicle semi-verticillate.

1. G. PAUCIFLORA Presl.—☉. San Francisco. Spring.

## 28. Poa L.

Spikelets 2-∞-flowered. Glumes 2, obtuse. Palets 2, both deciduous, lower carinate, upper 2-carinate. Ovary sessile.

1. P. DISTANS Grieseb. Lower palet rounded; all leaves short and narrow, mostly convolute.—♃. San Francisco. Summer.

2. P. CALIFORNICA Munro Ms. Diœcious; lower palet rounded; radical leaves half as long as the culm, mostly flat; culm-leaves short, frequently reduced to a mucro.—♃. San Francisco. Summer.

3. P. TENUIFOLIA Nutt. Lower palet rounded; radical leaves exceedingly narrow, mucronate; culm-leaves but little wider.—☉. San Francisco. Spring.

4. P. SCABRELLA Gray. Diœcious; lower palet rounded. leaves carinate.—♃. Oakland. Summer.

5. P. ANNUA L. Lower palet acute; branches of the panicle single or in pairs; all leaves flat. ☉. Common. Spring.

6. P. PRATENSIS L. Florets ☿; lower palet acute; branches of the panicle mostly in fives; the entire plant smooth, only the margins of the leaves slightly scabrous.—♃. Common. Summer.

7. P. TRIVIALIS L. Florets ☿: lower palet acute; branches of the panicle mostly in fives; the entire plant rough.—♃. Common. Summer.

8. P. Douglasii Nees. Dioecious; lower palet acute; radical leaves setaceously convolute.—♃. San Francisco. Sand dunes. Summer.

### 29. Eragrostis Beauv.

Spikelets 2-∞-flowered. Glumes 2, obtuse. Palets 2; upper palet persistent.

1. E. poæoides Beauv.—☉. San Francisco. Spring.

Probably introduced from Europe.

### 30. Briza L.  Quaking-grass.

Spikelets ∞-flowered; flowers imbricate, distichous. Glumes 2, concave, rounded, ventricose. Palets 2: lower rotund, concave, with cordate base; upper much smaller, 2-carinate. Caryopsis compressed. Leaves flattened. Spikelets pedicellate, paniculate.

1. B. media L.—♃. Common. Summer.

### 31. Festuca L.  Fescue Grass.

Spikelets 2-∞-flowered; flowers distichous. Glumes 2, carinate, unequal. Palets 2; lower not carinate, mucronate, sometimes aristate; upper 2-carinate. Ovary sessile, smooth.

1. F. myurus L. Monandrous; panicle contracted, spikelike, narrow, rays appressed; lower glume minute; upper glume half as long as the lowest floret.—☉. San Francisco. Spring.

2. F. TENELA Wild. Diandrous; panicle contracted, spikelike, lower rays in unequal pairs; glumes acute, the lower at least half as long as the upper.— ☉. San Francisco. Spring.

3. F. MICROSTACHYS Nutt. Monandrous; panicle contracted spike like; branches not appressed; glumes acute, the lower 1-nerved and smaller than the 3-nerved upper; upper glume almost as long as the lowest floret.--☉. Common. Spring.

4. F. OVINA L. Panicles loose, not spikelike; rays mostly solitary; glumes about equal; lower palet indistinctly nerved. — ♃. Coast Range. Summer.

5. F. SCABRELLA Torr. Panicles loose, not spikelike; lower rays distant in pairs; upper glume somewhat longer than the lower and half as long as the lowest floret; lower palet membranaceous, distinctly 5-nerved.—♃. Oakland. Summer.

6. F. PAUCIFLORA Thunb. Panicles loose, not spikelike; lower rays distant in pairs; glumes narrow, the upper twice as large as the lower; lower palet distinctly 5-nerved.—♃. San Francisco. Summer.

### 32. Bromus L.  BROME-GRASS.

Spikelets 3-∞-flowered; flowers distichous; glumes 2, unequal. Palets 2: lower rotund, convex, sometimes aristate; upper 2-carinate, the carina ciliate. Ovary sessile; hissute at the apex. Leaves flattened; spikelets pedicellate, paniculate.

1. B. MAXIMUS Desf. Lower glume 1-nerved, upper 3-nerved; glumes hyaline, the upper almost equalling the floret; lower palet scabrous.—☉. San Francisco. Mission. Spring.

2. B. SECALINUS L. Lower glume 3-5-nerved, upper 5-7-nerved; lower palet, inside rounded, convex, outside, carinate; rays of the panicle spreading, even when in fruit.—☉. Mission. Spring.

3. B. RACEMOSUS L. Lower glume 3-5-nerved, upper 5-7-nerved; lower palet, inside rounded, convex, outside, carinate; rays of the panicle contracted when in fruit.—☉ ☉. Oakland. Spring.

4. B. CILIATUS L. Lower glume 1-nerved, upper 3-nerved; glumes acute; the upper more than half the length of the lower floret; lower palet silky.—♃. San Francisco. Summer.

5. B. GRANDIFLORUS Hook (*Ceratochloa* Beauv.). Lower glume 5-nerved, the upper 9-nerved; lower palet compressed, carinate.— ♃. Nicasio. Summer.

### 33. Lepturus R. Br.

Spikelets 1-flowered, with rudiment of a superior flower. Glumes 2, rigid, subulate in the terminal spikelet opposite, in the lateral ones collateral; lower glume sometimes wanting. Palets shorter than the glumes, the lower enclosing the upper.—⊙. Leaves narrow and flattened. Spikelets solitary, immersed into alternate excavations of the rachis, the inflorescence representing a slender spike.

1. L. INCURVATUS Trin. Tiburon. Marin County. Spring.

Native of the Mediterranean region.

### 34. Lolium L. Darnel.

Spikelets ∞-flowered, flowers imbricate, distichous. Glumes 2, blunt, the posterior often wanting. Palets 2: the lower concave, sometimes aristate; upper 2-carinate with the caryopsis adhering. Leaves flattened. Spikelets solitary, immersed into alternate excavations of the rachis, placed edgewise with it, and representing a single spike.

1. L. PERENNE L. Glume much shorter than the spikelet.—♃. Common. Summer.

Native of Europe, but frequently escaped from cultivation.

2. L. TEMULENTUM L. Glume as long as the spikelet.—⊙. Common. Spring.

Has the reputation of being a narcotic poison, but varies considerably in regard to intensity and character of the symptoms caused by it.

### 35. Triticum L.  WHEAT.

Spikelets 3-∞-flowered. Flowers distichous, placed with their flat sides to the rachis. Glumes 2, lower sometimes wanting. Palets 2, lower rounded, upper 2-carinate, the carinate ciliate. Leaves flattened. Spikelets solitary, immersed into the excavations of the rachis, representing a single spike.

1. T. REPENS L.—♃. Common. Summer.

### 36. Hordeum L.  BARLEY.

Spikelets 1-flowered, with the rudiment of a superior floret, ternate; the central spikelet ☿, the lateral ones depauperate. Glumes 2, linear-lanceolate, aristate; all anterior. Palets 2; lower concave, aristate. Leaves flattened. Spikelets representing a single spike.

1. H. NODOSUM L. Glumes all setaceous from a broad base.—☉ ☉. Oakland. Summer.

2. H. MURINUM L. Glumes of the central spikelet lanceolate with a long awn and ciliate; outer glumes of the lateral spikelets setaceous; inner glumes like those of the central spikelet. ☉. San Francisco. Spring.

Native of Europe.

3. H. JUBATUM L. Glumes capillary, running into very long awns.—☉ ☉. Marin County. Summer.

"Squirrel-tail."

## 37. Elymus L.  Wild-Rye.

Spikelets 1-∞-flowered, all ☿ ; florets distichous, the uppermost rudimentary. Glumes 2, equal, anterior. Palets 2, low or concave. Ovary hirsute.—♃. Leaves flattened. Spikelets representing a single spike.

1. E. ARENARIUS L.  G l u m e s acuminate, shorter than the spikelet; lower palet carinate towards the cuspidate tip; leaves pungent.— Cliff House. Summer.

Introduced under the name "Esparto Grass," to keep the sand dunes from moving.

2. E. CONDENSATUS Presl. Glumes subulate-setaceous, shorter than the spikelet; lower palet 5-nerved, mucronate, sometimes shortly aristate; leaves ample, flat.—San Francisco. Summer.

3. E. SIBIRICUS L.  G l u m e s linear, 3-5-nerved, acute or shortly aristate; lower palet aristate, the awn longer than the palet.—San Francisco. Summer.

4. E. SITANION Schult. Glumes very long, sometimes 2-parted, ending in several very long awns.—San Rafael. Summer.

## 38. Gymnostichum Schreb.  Bottle-brush Grass.

Spikelets 1-4-flowered; florets remote, the uppermost rudimentary. Glumes rudimentary, often wanting. Palets 2, lower aristate, embracing the upper. — ♃. Leaves flattened. Spikelets representing a spike.

1. G. CALIFORNICUM Bolander.—Saucelito. Summer.

### Family 2. CYPERACEÆ.

Stem a calamus. Leaves all radical, sheathing but not split at their base; lamina parallel-veined. Anthers basifixed.

#### 1. Carex L.   SEDGE.

Flowers diclinic. ♂ spikelets 1-flowered; glume 1, external, ♀ spikelets 1-flowered with 2 glumes; the external like that of the ♂; the internal transformed into a utricle including the ovary.—♃. Leaves flattened. Stem triquetrous. Spikelets collected in diclinic or androgynous spikes.

1. C. MARCIDA Boott. Inflorescence the compound of many sessile spikes; spikes irregularly androgynous, sometimes the whole inflorescence ♀; stigmas 2; utricle (perigynium) ovate with a short 2-dentate apex; spikes almost black.—Santa Clara marshes. Summer.

2. C. DOUGLASII Boott. Diœcious; inflorescence the compound of many sessile spikes; utricle (perigynium) ovate, acuminate into a slender beak; spikes pale-brown. — Marin County. Summer.

3. C. MURICATA L. Inflorescence the compound of many sessile spikes; ♂ spikes or androgynous spikes at the top; inflorescence oblong, the compound of 4-10 spherical spikes; stigmas 2; bracts ovate, aristate, longer than the spikes; spikes chestnut-color.—Marin County. Summer.

4. C. GLOMERATA Thunb. Inflorescence the compound of many sessile spikes; ♂ or androgynous spikes at the top; inflorescence elongated, the compound of ∞-small rounded spikes; stigmas 2; bracts setaceous, longer than the spikes; spikes brown. —Common. Summer.

5. C. PANICULATA L. Inflorescence the compound of many sessile spikes; ♂ or androgynous spikes on the top; inflorescence almost linear, the compound of oblong spikes (sometimes branched); stigmas 2; bracts setaceous; spikes brown or pale.—Presidio. Summer.

6. C. FESTIVA Dewey. Inflorescence the compound of many sessile spikes, forming a capitulum; ♂ spikes at the bottom or centre; spikes brown.—Oakland. Summer.

7. C. DEWEYANA Schwein. Spikes of the inflorescence distinct, androgynous; ♂ florets at the base; stigmas 2; spikes pale. — Napa Valley. Summer.

8. C. BIFIDA Boott. Spikes of inflorescence distinct, the uppermost androgynous with ♂ florets on the base, the rest of the spikes ♀,

stigmas 3; spikes dark-colored.—Marin county. Summer.

9. C. NUDATA Bolander. Spikes unisexual, the upper ♂, the rest ♀ (sometimes androgynous) stigmas 2; ♂ spike single (sometimes androgynous); utricle (perigynium) elliptic, compressed, slenderly nerved.—Tamalpais. Summer.

10. C. AQUATILIS Wahl. Spikes unisexual; the upper 1-4 spikes ♂, the lower (2-5) ♀; stigmas 2; utricle (perigynium) elliptic, stipitate, nerveless.—Wright's Station. Summer.

11. C. SITCHENSIS Prescott. Spikes unisexual; the upper 1-4 spikes ♂, the rest (3-5) ♀; stigmas 2, utricle (perigynium) orbicular, turgid, stipitate, coriaceous, nerveless.—Presidio. Summer.

12. C. JAMESII Torr. Spikes unisexual; upper 1-4 spikes ♂, the rest (3-4) ♀; stigmas 2, utricle (perigynium) oval, strongly nerved, abruptly ending in a 2-dentate beak; bracts about the length of the stem.—Sonoma. Summer.

13. C. LACINIATA Boott. Spikes unisexual; upper 1-2; spikes ♂, the rest (3-4) ♀; stigmas 2, utricle (perigynium) oval, lenticular compressed, bracts far exceeding the stem.—Coast Range. Summer.

14. C. GLOBOSA Boott. Spikes unisexual, upper spike ♂, rest (3-5) ♀; stigmas 3; utricle

(perigynium) globose, tapering to the base, abruptly rostrate; beak short.—Marin county. Summer.

15. C. PSEUDOCYPERUS L. Spikes unisexual, upper 1 ♂ ; rest (3–5) ♀ , drooping; stigmas 3; utricle coriaceous, ovate, triquetrous, attenuate into a long slender beak.—San Francisco. Summer.

## 2. Fimbristylis Vahl.

Spikelets ☿, ∞-flowered; palets closely imbricate all round, the lowermost empty. Perigonium 0. Style incrassate at base, persistent. Leaves narrow. Spikelets, if solitary, bracteate; if capitate or umbellate, involucrate.

1. F. MILIACEA Vahl.—⊙. San Francisco. Folsom street. Spring.

This tropical species has not been found again since the filling in of that part of the city.

## 3. Isolepis L.

Spikelets ☿, ∞-flowered. Palets imbricate all round, the lowermost empty. Perigonium 0. Base of the style deciduous, cespitose, slender.

1. I. RIPARIA R. Br. Nutlets triquetrous, the sides convex.—⊙. Common. Spring.

2. I. CARINATA Hook. & Arn. Nutlets triquetrous, the sides straight.—⊙. San Francisco. Spring.

### 4. Scirpus L.  BULRUSH.

Spikelets ☿, ∞-flowered. Palets imbricate all round, the lowermost empty. Perigonium 6-setaceous. Style articulate with its base.

1. S. LACUSTRIS L. Inflorescence apparently lateral, with a single erect involucral leaf; stem almost terete; bristles of perigonium 6, slender, with scattered barbs—♃. Common. Summer.

2. S. TATORA Kunth. Inflorescence apparently lateral, with a single erect involucral leaf; stem almost terete; bristles of perigonium less than 6, stout and retrorsely plumose.—♃. Benicia. Summer.

3. S. OLNEYI Gray. Inflorescence apparently lateral, with a single erect involucral leaf; stem triquetrous.—♃. Presidio. Summer.

4. S. MARITIMUS L. Involucre spreading; spikelets large in a sessile cluster or sparingly umbellate.—♃. Common. Summer.

5. S. SYLVATICUS L. Involucre spreading; spikelets small in a supra-decompound inflorescence.—♃. Common. Summer.

6. S. ACICULARIS L. (*Eleocharis* R. Br.). Spike distichous or 3-ranked, few-flowered, terminating a leafless stem.—☉. Common. Spring.

7. S. PALUSTRIS Reichenb. (*Eleocharis* R. Br.). Spike terete, ∞-flowered, terminating a leafless stem.—♃. Common. Summer.

### 5. Eriophorum L. COTTON-GRASS.

Spikelets ☿, ∞-flowered. Palets imbricate all round, the lowermost empty. Perigonium ∞-setaceous, in fruit exserted, and silky.—♃.

1. E. GRACILE Koch.—San Francisco. Summer.

Aquatic; formerly found in a swamp near the mouth of Mission creek. Not yet rediscovered in our neighborhood, but frequent in the higher Sierras.

### 6. Cyperus L.

Spikelets ☿. Palets imbricate, distichous, the lowermost empty. Perigonium 0. Inflorescence involucrate.

1. C. DIANDRUS Torr. Style 2-cleft; nutlet lenticular, the edge turned to the rachis of the spikelet.—☉. San Francisco. Summer.

2. C. ARISTATUS Rottl. Style 3-cleft; nutlet triquetrous. Floret monandrous.—♃. Common. Spring.

ORDER 5. SPADICIFLORÆ. Inflorescence a spadix surrounded by a spathe. Bracts depauperate. Ovary superior, one to several-celled; cells 1-ovulate. Leaves alternate with sheathing base.

### Family 1. TYPHACEÆ.

Flowers diclinous, in a sometimes interrupted club-shaped spadix, each division of which is protected by foliaceous spathes. Upper part of spadix ♂; perigonium 0; anthers basifixed, irregularly mixed with scales. Lower portion of spadix ♀; ovaries surrounded at their bases by ∞-bristles, sometimes by 3 scales; stigma 1, lateral. Fruit a utricle; endosperm copious. Leaves linear, straight-nerved. Aquatic.—♃.

### 1. **Sparganium** Tourn.   Bur-Reed.

Flowers in crowded glomerulate heads, constituting a simple or a branched spadix. Filaments slender. Ovaries sessile, surrounded by some linear scales.

1. S. EURYCARPUM Engelm.—West Oakland (extinct); rediscovered at Niles. Summer.

### 2. **Typha** Tourn.   Cat-Tail. Flag.

Flowers in a club-shaped, continuous spadix. ♂: filaments connate, very short. ♀: ovaries stipitate, surrounded by ∞ bristles.

1. T. LATIFOLIA L.—Common. Summer.

Series 3. HELOBIÆ. Aquatics. Type generally ternate. Seed nearly without endosperm. Embryo with very pronounced radicle.

Order 1. POLYCARPICÆ. Ovaries 3-∞. Endosperm entirely wanting.

### Family 1. ALISMACEÆ.

Calyx 3-sepalous. Corolla 3-petalous. Stamens 6-∞. Ovaries 3-∞, with terminal styles and ventral placentation. Carpidia dry, indehiscent. Lamina of leaves nervose.

### 1. Alisma L.  WATER-PLANTAIN.

Flowers ☿. Calyx herbaceous. Corolla petaloid. Ovaries ∞, 1-ovulate. Akenes ∞, verticillate.—♃. Aquatic. Scapigerous. Inflorescence verticillate, paniculate.

1. A. PLANTAGO L. Flowers pale.—Common. Summer.

### Family 2. JUNCAGINEÆ.

Calyx not differing from corolla. Lamina of leaves abortive.

### 1. Triglochin L.  ARROW-GRASS.

Perigonium deciduous, 3 inner sepals sometimes wanting. Ovary 6-celled; ovules in each cell solitary, basal. Styles 3-6; stigmas plumose. Capsule separating into carpidia; finally opening ventrally.—♃. Scapigerous. Inflorescence spicate.

1. T. MARITIMUM L. Common. Summer.

### Family 3. POTAMEÆ.

Perigonium if present, 4-cleft. Stamens 1, 2, or 4. Ovaries 4-∞, 1-ovulate. Carpidia indehiscent. Leaves stipulate.

### 1. Potamogeton TOURN.  PONDWEED.

Flowers ☿. Perigonium 4-cleft. Anthers 4, sessile, inserted on the base of the sepals.

Ovaries 4, styles 0.—♃. Jointed aquatics. Inflorescence pedunculate; spikes axillary.

1. P. NATANS L. Floating leaves coriaceous; submerged leaves filiform. — Marine Hospital. Lake Merced. Taylorville. Summer.

2. P. LUCENS L. Leaves uniform, lanceolate.—Mission Dolores. Lake Merced. Summer.

3. P. PAUCIFLORUS Pursh. Leaves linear with distinct stipules.—Ocean Lake. Summer.

4. P. PECTINATUS L. Leaves linear, stipules united with the sheathing base of the leaf.—Common. Summer.

ORDER 2. CENTROSPERMÆ. Perigonium 0. Ovules basilar.

### Family 1. NAIADACEÆ.

Flowers diclinous. Ovary 1, 1-celled, 1-ovulate. Fruit a nutlet.

#### 1. Lilæa HBK.

Flowers monœcious. ♂, spicate, 1-bracteate, monandrous, sometimes mingled with ♀ flowers. ♀ flowers partly spicate and 1-bracteate; partly solitary, axillary and without bracts. Styles of the upper ♀ flowers short, of the lower elongate; stigmas capitate.—☉. Aquatic. Leaves radical, terete, sheathing at base.

1. L. SUBULATA HBK.—San Francisco. School House Station. Searsville. Spring.

## 2. Zostera L.  EEL-GRASS.

Flowers monœcious. Spathe elongated into a lamina. Spadix flattened, dorsally naked, ventrally covered by alternate stamens and ovaries arranged in two rows. Anthers 1-celled, sessile. Ovary rostrate. Style persistent.— ♃. Stems and roots submerged.

1. Z. MARINA L.—Brackish inlets around the bay. Summer.

The bulbous root is edible, and is eagerly sought for by the canvas-back duck.

## Family 2. LEMNACEÆ.

Aquatic plants without axis. Flowers enclosed in a thin membranaceous spathe. Stamens 1 or 2. Ovary sessile, 1-celled; ovules 1-6; style short. Fruit a utricle.

### 1. Lemna L.  DUCKWEED.

Stamens 2.

1. L. TRISULCA L. Ovule solitary; fronds oblong, stalked, remaining connected.—Presidio. Autumn.

2. L. VALDIVIANA Philippi. Ovule solitary; fronds not stalked, soon separating; bract reniform.—Marine Hospital. Autumn.

3. L. MINOR L. Ovule solitary; fronds not stalked, soon separating; bract cucullate.—Outlet of a creek between the Fort and the Cliff House. Autumn.

4. L. GIBBA L. More than 1 ovule.—Common. Autumn.

Sub-Class 2. GYMNOSPERMÆ.

### Order 1. CONIFERÆ.

#### Family 1. TAXINEÆ.

##### 1. **Torreya** Arn. NUTMEG-PINE.

Flowers from scaly buds. ♂ : consisting of bracts on the base of the axis, imbricated in 4 rows; stamens ∞; connectives peltate; anthers 4 to each stamen. ♀ : ovule immersed into an urceolate arillus.—♃. Branches opposite. Leaves linear, decurrent, rigid, mucronate.

1. T. CALIFORNICA Torr.—Tamalpais. Lagunitas Creek. Spring.

Wood and seed have the odor of nutmeg, the seed occasionally being used as such.

#### Family 2. ABIETINEÆ.

##### 1. **Pinus** Tourn. PINE.

♂ flowers on elongated axes; the different axes crowded into an inflorescence round the base of a new shoot. ♀ flowers with fertile scales protected by bracts, which are much smaller than the scales. Scales of the cone persistent and elevated into a tumor (umbo).— ♃. Leaves 1-5, from a squamous sheath.

1. P. SABINIANA Dougl. Leaves in threes; cones on a well developed peduncle; apophyses of the scales stout, projecting.—Napa Valley. Spring.

"Blue-pine." Seeds eaten by the Indians.

2. P. INSIGNIS Dougl. Leaves in threes; cones on short peduncles in clusters. Scales toward the base enlarged, thick and hemispherical.—♃. Bolinas heights. Santa Cruz mountains. Spring.

"Monterey pine." Frequently cultivated.

3. P. TUBERCULATA Gordon. Leaves in threes; cones verticillate, peduncled; scales angular, enlarged by a conical umbo.—Marin County. Spring.

4. P. MURICATA Don. Leaves in pairs. Bolinas. Spring.

### 2. Pseudotsuga Carriere. DOUGLAS SPRUCE.

Flowers from the axils of last year's leaves, crowded; ♂ on elongated axes; scales of ♀ flower considerably smaller than the bracts, persistent. Cones pendulous.— ♄. Leaves distichous, petioled.

1. P. DOUGLASII Carr.—Tamalpais. Santa Cruz mountains. Spring.

### Family 3. CUPRESSINEÆ.

#### 1. Sequoia Endl.

Cones ovate; scales $\infty$, cuneiform, spreading at maturity, decussately arranged.— ♄. Leaves alternate, decurrent, carinate.

1. S. SEMPERVIRENS Endl. Leaves spreading, distichous. Branchlets spreading.—Marin County. Coast Range. Contra Costa Range. Spring. "Redwood."

2. S. GIGANTEA Decaisne. Leaves not distichous. Branchlets pendulous.—In a gulch of the Santa Cruz mountains near Meyers' vineyard. Otherwise only in the high Sierras.
"Big tree."

#### 2. Cupressus Tourn.   CYPRESS.

Cone globose; scales 6-10, thick, peltate, valvate, $\infty$-ovulate. Seeds angulate, narrowly winged.— ♄. Leaves decussate, adnate, squamate, imbricate, not distichous.

1. C. MACROCARPA Hartw. Cones of 5 or 6 pairs of scales.—Monterey. In our local flora only escaped from cultivation. Spring.
Monterey Cypress.

2. C. GOVENIANA Gordon. Cones of 3 or 4 pairs of scales.—Nicasio. Spring.

#### 3. Juniperus L.   JUNIPER.

Fruit a galbulus. Scales succulent, uniting in fruit. Seeds osseous.— ♄. Leaves opposite, not distichous.

1. J. CALIFORNICA Carr.—Walnut Creek. Spring.

## CLASS 2. CRYPTOGAMÆ.

Section 1. DICHOTOMÆ (*Lycopodiaceæ*). Ramification dichotomous. Each leaf producing only one sporangium, borne on the upper surface of the leaf or near its axil.

### Family 1. SELAGINELLEÆ.

Macrosporangia and Microsporangia. Leaves small and of two different kinds.

#### 1. Selaginella Beauv.

Sporangia axillary, globose, transversely dehiscent.—♃. Leaves 4-8-ranked.

1. S. RUPESTRIS.—Spring. Tamalpais.

### Family 2. ISOETEÆ.

Heterosporous. All leaves long and grass-like.

#### 1. Isoetes L. QUILLWORT.

Characters of the family of which it is the only genus known.

1. *sp.* Submerged aquatic.—Corte Madera. Found by Mrs. Curran, but not yet identified.

2. *sp.* Growing on places inundated only during the rainy season.—Olema.
Found by Mrs. Curran, but not yet identified.

Section 2. FILICINEÆ. Ramification not dichotomous. Leaves developed, frondose. Sporangia not single and never on the upper surface of the leaf.

ORDER 1. RHIZOCARPÆ. Sporangia in sporocarps (conceptacula); heterosporus; microsporangia, ∞-spored; macrosporangia generally 1-spored.

### Family 1. MARSILÆACEÆ.

Sporocarps 2–4-celled, 2–4-valved, containing macro- and microsporangia. Sporangia parietal; macrosporangia, always 1-s p o r e d. Vernation circinate.

#### 1. Marsilæa L.

Macro- and microsporangia in the same sporocarp.—♃. Fronds petiolate, 4-foliolate.

1. M. VESTITA Hook. & Grev.—Fort Point. (Extinct.)

#### 2. Pillularia L.

♃. Fronds filiform.

1. P. AMERICANA Al. Braun.—Marin County.

### Family 2. SALVINIACEÆ.

S p o r o c a r p s 2 to ∞, on the same petiole (transformed half of a frond), macro- and microsporangia in distinct sporocarps. Vernation not circinate.

### 1. Azolla Lam.   WATER-FERN.

Sporocarps in pairs. Macrosporangia 1-spored. Aquatics, moss-like, floating, green or purplish.

1. A. CAROLINIANA Willd.—Presidio. Saucelito.

ORDER 2. FILICES. Isosporous. Sporangia formed from the epidermis, bursting by an elastic ring. Fronds without stipules. Vernation circinate.

### Family 1. PÓLYPODIACEÆ.

Sporangia hypophyllous, somewhat globular; ring vertical.

#### 1. Gymnogramme Desf.   GOLD-FERN.

Sori linear, placed on branching primary veins. Indusium 0.—♃.

1. G. TRIANGULARIS Kaulfuss.—Common.

#### 2. Cheilanthes Swartz.   LIP-FERN.

Sori placed separately at the end of veinlets. Spurious indusium rudimentary, formed by the reflexed margin of the frond.—♃.

1. C. CALIFORNICA Nutt. Fronds green on both sides, 4-pinnatifid.—Tamalpais.

2. C. MYRIOPHYLLA Desv. Fronds covered underneath with brown scales; 3-4-pinnate.—Wright's Station.

### 3. Allosorus Benth. (*Pellæa* Link.)　Rock-Brake.

Sporangia on veins which are oblique to the midrib. Spurious indusium continuous, formed by the reflexed margin of the frond.—♃. Glabrous, cespitose; fertile fronds contracted; petioles not green.

   1.  A. ANDROMEDÆFOLIUS Fée. Pinnules oval, obtuse.—San Rafael. Tamalpais.

   2.  A. ORNITHOPUS Hook. Pinnules rhomboid, mucronulate.—Piedmont. Lagunitas reservoir.

   3.  A. DENSUS Hook. Pinnules linear, mucronate.—Lagunitas reservior.

### 4. Polypodium L.　Polypody.

Sori round, placed on the veins. Indusium 0.—♃.

   1.  P. VULGARE L. Veins of the segments uniformly free and none anastomosing.—Common.

   2.  P. SCOULERI Hook & Gray. Veins of the segments anastomosing by veinlets.—Saucelito. Tamalpais.

### 5 Pteris L.　Bracken. Brake.

Sporangia at the end of the veinlets, connected into a vein-like receptacle which surrounds the frond without interruption, indusium continuous, membranaceous, attached at its marginal side, free on its inner side.—♃.

1. P. AQUILINA L. Common. The fronds before expanding are edible, and were used in former times as a pot herb.

### 6. Adiantum L.
MAIDEN'S HAIR. FIVE-FINGER.

Sporangia at the end of distinct veins. Sori covered by the semi-lunar reflexed margins. Indusium continuous with the margin of the leaf, free on its inner side.—♃. Petioles not green.

1. A. EMARGINATUM Hook. Rachis of the frond continuous to the apical pinnule (leaflet). Common.

2. A. PEDATUM L. Rachis of the frond 2-furcate, each partition bearing several pinnate branches on its apical side.—Camp Taylor. Saucelito.

### 7. Blechnum L. (*Lomaria*, Willd.) DEER-FERN.

Sporangia forming a linear sorus on each side of the mid-rib and parallel to it. Indusium membranaceous, attached to the receptacle and free on the inner side.—♃.

1. B. SPICANT Desvaux.—Tennessee Valley. Wild-wood Glen.

### 8. Woodwardia L. CHAIN-FERN.

Sporangia placed on reticulate veinlets, forming oblong sori on each side of the mid-rib. Indusium coriaceous, attached to the receptacle, and opening at the inner side.—♃.

1. W. RADICANS Smith.—Common.

### 9. Asplenium L.   SPLENWORT.

Sori linear, placed on veinlets. Indusium membranaceous, attached laterally to the veinlet and opening toward the mid-rib.—♃.

1. A. FILIX FŒMINA Bernh.—Marine Hospital. Tamalpais. "Lady-fern."

### 10. Aspidium Swartz.   SHIELD-FERN.

Sori round, placed at the end of the veinlets. Indusium umbilicate, free on all sides and affixed by a column.—♃.

1. A. RIGIDUM Swartz. Indusium reniform; fronds 2-pinnate.—Common.

Specific against the tapeworm, like its European congener, *A. Filix mas.*

2. A. MUNITUM Kaulfuss. Indusium peltate; fronds pinnate, pinnæ serrate.—Common.

3. A. ACULEATUM Swartz. Indusium peltate; fronds 2-pinnate, sometimes simply pinnate, the pinnæ deeply cut.—Tamalpais.

### 2. Cystopteris Bernh.   BLADDER-FERN.

Sori round, placed on the middle of the veinlet, solitary in the disc of the lacinia. Indusium hyaline, laterally affixed, free towards the margin of the frond.— ♃.

1. C. FRAGILIS Bernh.—Tamalpais.

ORDER 3 STIPULATÆ. Isosporous. Leaves stipulate.

### Family 1. OPHIOGLOSSEÆ.

Sporangia capsular developed in the parenchyma of half of the frond. Vernation not circinate.

#### 1. Botrychium Swartz.

Fertile segment of the frond a compound spike. Sporangia distinct, glabrous, distichous, nearly 2-valved.— ♃.

1. B. TERNATUM Swartz.—San Francisco. (Extinct.)

Section 3. EQUISETACEÆ. Stem articulate; internodes surrounded by sheathing whorls of scarious leaves. Sporangia on the under side of verticillate, peltate receptacles forming a terminal spike. Spores $\infty$, each furnished with two elaters.

#### 1. Equisetum L.
HORSETAIL. SCOURING-RUSH.

Only genus.

1. E. TELMATEIA Ehrh. Sterile stems green with verticillate branches; fertile stems with-

out chlorophyll, unbranched.—Marin County. Contra Costa. Spring.

2. E. ROBUSTUM Al. Braun. All stems green, unbranched. — Fort Point. Marin County. Spring.

# INDEX.

ABIETINEÆ..............326
Abronia............25, 200
   latifolia...............200
   umbellata.............200
Acæna, 5, 6, 7, 8, 17, 25, 38
   40, 41, 250.
   trifida................250
Acanthomintha....7, 51, 136
   lanceolata............136
Acer........23, 36, 74, 211
   macrophyllum....... 211
Achillea............67, 108
   millefolium...........108
Achyrachæna....... 67, 102
   mollis................102
Actæa.............48, 233
   spicata...............233
Adenocaulon.......... 88
   bicolor................ 88
Adenostoma......40, 44, 250
   fasciculatum..........250
Adenostyles............ 60
Adiantum..............333
   emarginatum..........333
   pedatum..............333
Æsculus.........35, 74, 211
   Californica............211
Agrostis............12, 300
   alba..................300
   exarata...............301
   scabra................301
   verticillata............301

Aira...............13, 304
   cæspitosa............ 304
   danthonoides.........305
   elongata..............305
   holciformis...........304
Alchemilla..........35, 250
   arvensis..............250
Alfilerilla..............205
Alder..................271
Alisma.............35, 323
   plantago..............323
ALISMACEÆ.............322
Allium.............33, 286
   attenuifolium........287
   lacunosum...........287
   serratum.............287
   unifolium............287
Allosorus..............332
   andromedæfolius......332
   densus...............332
   ornithopus...........332
Alnus.............74, 271
   rhombifolia..........272
   rubra................272
Alopecurus.........11, 298
   geniculatus..........298
   pratensis.............298
*Alsine*.............. 30, 42
ALSINEÆ...............192
Alum-root.............187
Alyssum............54, 227
   calycinum........... 227

Alyssum—maritimum...228
AMARANTACEÆ..........197
Amarantus....75, 77, 79, 197
   retroflexus..........197
Ambrosia.............75, 92
   artemisiæfolia.........92
   psilostachya..........92
AMBROSIÆ.............92
Amelanchier........45, 252
   alnifolia..............253
Amorpha.........56, 57, 262
   Californica..........262
AMPELIDEÆ.............214
Amsinckia..........19, 155
   intermedia...........155
   lycopsoides..........156
   spectabilis..........155
   tesselata.............155
AMYGDALACEÆ..........245
ANACARDIACEÆ.........208
Anagallis..........20, 172
   arvensis.............172
Anaphalis...........65, 90
   margaritacea..........90
Anemone............49, 237
   Grayi...............237
Anemopsis......35, 38, 279
   Californica...........279
Angelica...........28, 178
   tomentosa...........178
ANGIOSPERMÆ...........80
ANISOCARPÆ............80
Anoplanthus........51, 142
   fasciculatus.........143
   uniflorus...........142
ANTHEMIDEÆ...........107
Anthemis...........67, 107

Anthemis—cotula.......107
Anthoxanthum......11, 299
   odoratum............300
ANTIRRHINEÆ..........151
Antirrhinum........53, 152
   vagans..............153
Aphyllon.......51, 142, 143
   comosum............143
APHANOCYCLICÆ.........221
Apiastrum............181
   angustifolium........181
Apium.................27
Aplopappus......61, 63, 84
   ericoides.............84
   linearifolius..........84
APOCYNACEÆ...........129
Apocynum.........25, 129
   androsæmifolium.....129
   cannabinum.........129
Aquilegia..........49, 234
   truncata............234
Arabis............55, 226
   blepharophylla.......227
   perfoliata...........227
Aralia.............31, 184
   Californica..........184
ARALIACEÆ............184
Arbutus............41, 174
   Menziesii...........175
Arceuthobium.....77, 267
   occidentale..........267
Arctostaphylos....41, 174
   Andersonii..........174
   pungens............174
   tomentosa..........174
Arenaria..........42, 193
   Californica..........193

| | |
|---|---|
| Arenaria—Douglasii ....193 | Aster—radulinus........ 86 |
| macrophylla .........193 | ASTEROIDEÆ............ 81 |
| palustris................193 | Astragalus ..........58, 263 |
| Argemone.......47, 48, 230 | crotalariæ ............263 |
| hispida ........... 230 | didymocarpus ........263 |
| Aristolochia......... 71, 268 | Douglasii.............263 |
| Californica ...........268 | leucophyllus..........263 |
| ARISTOLOCHIÆ...........268 | Menziesii.............263 |
| Armeria..........31, 170 | pycnostachyus........264 |
| vulgaris ..............170 | tener................263 |
| Arnica ..............65, 111 | Atriplex......75, 77, 79, 198 |
| discoidea..................111 | Californicum ........199 |
| Arrhenatherum......12, 305 | coronatum............198 |
| avenaceum...........305 | leucophyllum.........199 |
| Arrow-grass ........... 323 | patulum................198 |
| Artemisia .......61, 69, 109 | Audibertia ...........7, 137 |
| Californica ...........110 | grandiflora ...........137 |
| dracunculoides .......110 | stachyoides...........137 |
| Ludoviciana........ 110 | Avena ..............13, 303 |
| pycnocephala.........110 | Azolla................331 |
| Asarum ............43, 268 | |
| caudatum .......... 268 | Baccharis .....61, 62, 67, 87 |
| ASCLEPIADACEÆ.........128 | Douglasii.............87 |
| Asclepias ..............128 | pilularis ............87 |
| Mexicana ............129 | viminea............87 |
| speciosa..............128 | Bœria ............64, 103 |
| vestita...............128 | carnosa .............104 |
| Ash...................132 | chrysostoma.........104 |
| Aspidium ..........234 | Fremontii ...........104 |
| aculeatum............234 | gracilis............. 104 |
| munitum ............234 | macrantha...........103 |
| rigidum..............234 | uliginosa .............104 |
| Asplenium ............334 | *Bahia* ................104 |
| Filix fœmina........ 334 | Balsamorrhiza ..... 66, 94 |
| Aster...............63, 86 | deltoidea..............94 |
| Chamissonis .........86 | Hookeri............. 94 |
| divaricatus ........ 86 | Baneberry........... 233 |

| | | | |
|---|---|---|---|
| Barbarea | 55, 224 | Blue-Curls | 132 |
| vulgaris | 224 | Blue-eyed Grass | 282 |
| Barberry | 232 | Blue-pine | 326 |
| Barley | 314 | Boisduvalia | 37, 244 |
| Bay-tree | 231 | cleistogama | 244 |
| Beard-grass | 300 | densiflora | 244 |
| Beckmannia | 298 | glabella | 244 |
| erucæformis | 298 | BORRAGINEÆ | 155 |
| Bed-straw | 126 | Botrychium | 335 |
| Beggar-ticks | 96 | ternatum | 335 |
| Bellflower | 122 | Bottle brush grass | 315 |
| Bent-grass | 300 | Bowlesia | 27, 183 |
| BERBERIDEÆ | 231 | lobata | 184 |
| Berberis | 31, 232 | Boykinia | 30, 186 |
| aquifolium | 232 | occidentalis | 186 |
| nervosa | 232 | Box-Elder | 211 |
| pinnata | 232 | Brass-buttons | 109 |
| repens | 232 | Bracken | 332 |
| Berula | 28, 180 | Brake | 332 |
| angustifolia | 180 | Brassica | 56, 225 |
| BETULACEÆ | 271 | campestris | 225 |
| Bidens | 62, 68, 96 | nigra | 225 |
| chrysanthemoides | 96 | *Brickellia* | 112 |
| Big-root | 122 | Bridal-wreath | 247 |
| Big-tree | 328 | Briza | 14, 310 |
| Bigelovia | 60, 63, 84 | media | 310 |
| arborescens | 84 | Brizopyrum | 77, 307 |
| Bind-weed | 168 | maritimum | 308 |
| Bitter-Cress | 227 | Brodiæa | 8, 32, 288 |
| Blackberry | 247 | congesta | 288 |
| Bladder-fern | 334 | grandiflora | 288 |
| Bladderwort | 170 | minor | 288 |
| Blechnum | 333 | terrestris | 288 |
| spicant | 333 | Brome-grass | 311 |
| Bleeding-heart | 228 | Bromus | 14, 311 |
| Blennosperma | 69, 106 | ciliatus | 312 |
| Californicum | 107 | grandiflorus | 312 |

| | |
|---|---|
| Bromus—maximus ......312 | CALYCIFLORAE...........238 |
| racemosus...........312 | Camassia............33, 287 |
| secalinus.............312 | Campanula......... 22, 122 |
| Brooklime..............148 | exigua............ 122 |
| Brookweed...............172 | prenanthoides........122 |
| Buck-bean..............131 | CAMPANULACEÆ..........121 |
| Buckeye................211 | Campion................191 |
| Bulbostylis.........60, 112 | Canary-grass.......... .. 298 |
| Californica...........112 | Canchalagua............130 |
| Bulrush................320 | CAPRIFOLIACEÆ..........124 |
| Bur-reed..............322 | Capsella.............54, 223 |
| Burrielia............64, 103 | Bursa-pastoris........223 |
| microglossa...........103 | Caraway................181 |
| Buttercup..............235 | Cardamine..........54, 227 |
| Button-bush........ ...126 | oligosperma ..........227 |
| | paucisecta............227 |
| Cakile .... .......54, 224 | Carex..............72, 316 |
| Americana............224 | aquatilis..............318 |
| Calais..................116 | bifida ................317 |
| Kelloggii..............117 | Deweyana............317 |
| Lindleyi.......... 117 | Douglasii.... ........316 |
| linearifolia...........117 | festiva................317 |
| *Calamagrostis*........12, 302 | globosa. .............318 |
| Calandrinia, 30, 35, 42, 44 | glomerata............317 |
| 45, 195. | Jamesii...............318 |
| Menziesii.... ... ....195 | laciniata..............318 |
| California Lilac..........213 | marcida ..............316 |
| California Poppy........229 | muricata............317 |
| Calochortus... ......33, 290 | nudata ............. 318 |
| albus............... 291 | paniculata ...........317 |
| lilacinus.............291 | pseudocyperus........319 |
| luteus................291 | Sitchensis ...........318 |
| pulchellus...........291 | Carrot..................176 |
| venustus ...... ......291 | Carthamus............. 62 |
| CALYCANTHACEÆ........252 | Carum.............28, 181 |
| Calycanthus.........46, 252 | Kelloggii.. . .........181 |
| occidentalis...........252 | CARYOPHYLLALES........190 |

15A

| | |
|---|---|
| Cascara sagrada........ .. 212 | Ceratophyllum — demersum................280 |
| Castanopsis .........76, 270 | |
| chrysophylla.........271 | Cercis. ............39, 265 |
| Castilleia........... 52, 147 | occidentalis ........265 |
| affinis...............147 | Chænactis..........61, 106 |
| foliolosa.. ...........147 | lanosa..............106 |
| latifolia......... ....147 | Chain-fern .............333 |
| parviflora..... ... .. 147 | Chamomile............108 |
| Catchfly...............191 | Chapparal............. 213 |
| Caterpillar-grass ...... .298 | Cheilanthes............331 |
| Cat-tail................322 | Californica ........ 331 |
| Caucalis ............29, 175 | myriophylla .........331 |
| microcarpa....... ...176 | Cheiranthus........ 55, 226 |
| nodosa..............175 | Chemisal..............250 |
| Caulanthus..... ....55, 225 | CHENOPODIACEÆ.........197 |
| procerus.............225 | CHENOPODIALES.........197 |
| Ceanothus ..........23, 213 | Chenopodium 5, 7, 8, 25, 197 |
| crassifolius .......... 214 | album ..............197 |
| cuneatus ............214 | ambrosioides .........198 |
| dentatus ............213 | Californicum ........198 |
| divaricatus...........213 | murale..............198 |
| papillosus ...........214 | Cherry ................246 |
| thyrsiflorus ...... 213 | Chickweed . ..........192 |
| CELASTRACEÆ............214 | Chinquapin............. 27 |
| CELASTRALES............212 | Chlorogalum........34, 286 |
| Centaurea.......62, 68, 114 | pomeridianum........286 |
| benedicta..............114 | Chorizanthe..15, 34, 39, 277 |
| melitensis........... 114 | Douglasii............278 |
| CENTROSPERMÆ.....190, 234 | membranacea.........278 |
| Cephalanthus...........126 | pungens............. 278 |
| occidentalis .........126 | Chrysanthemum........108 |
| Cerastium...........43, 192 | segetum..............108 |
| arvense..............192 | Chrysopsis........61, 63, 84 |
| pilosum..............192 | Oregana ............84 |
| *Ceratochloa*.............312 | sessiliflora ............84 |
| CERATOPHYLLEÆ ........279 | *Cicendia* .. .........16, 130 |
| Ceratophyllum ......75, 280 | CICHORIACEÆ ...........115 |

| | |
|---|---|
| Cicuta..............28, 180 | Collomia ...........21, 167 |
| Bolanderi ............180 | gilioides .............168 |
| Californica ..........180 | gracilis...............167 |
| maculata.............180 | heterophylla..........168 |
| CISTINEÆ...............217 | Columbine .............234 |
| Clarkia............. 37, 243 | COMPOSITÆ............. 80 |
| elegans ..............244 | CONIFERÆ..............326 |
| Claytonia...........30, 195 | Conium..............29, 181 |
| Chamissonis..........196 | maculatum............182 |
| linearis ..............196 | CONVOLVULACEÆ........168 |
| parviflora............196 | Convolvulus ........21, 168 |
| perfoliata ...........196 | arvensis..............169 |
| Sibirica ..............195 | Californicus ..........169 |
| spathulata...........196 | luteolus..............169 |
| Cleavers .................126 | pentapetaloides.......169 |
| Clematis.............49, 237 | Soldanella............168 |
| lasiantha.............237 | Coral-root. ............281 |
| ligusticifolia..........237 | Corallorhiza.........71, 281 |
| Clintonia ...........33, 285 | Bigelovii .... ........282 |
| Andrewsiana .........285 | multiflora ............282 |
| Clover ..................256 | Cordylanthus.....7, 52, 144 |
| Cnicus .............62, 112 | filifolius..............144 |
| Americanus...........113 | maritimus...... .... 144 |
| Breweri..............114 | mollis................144 |
| edulis .... ..........113 | pilosus...............144 |
| fontinalis.............113 | Corethrogyne.........68, 85 |
| Hallii ................113 | obovata .............. 85 |
| occidentalis.... ......113 | Corn-spurrey ..........194 |
| quercetorum..........113 | CORNACEÆ.... ........185 |
| Cocklebur............... 93 | Cornel ................185 |
| Coinogyne ..........63, 103 | CORNICULATÆ......... 186 |
| carnosa ..............103 | Cornus...............17, 185 |
| Collinsia............53, 152 | Californica ..........185 |
| bartsiæfolia...........152 | Nuttallii..... .......185 |
| bicolor ......... ....152 | CORONARIÆ.............283 |
| parviflora........... .152 | Corylus.............76, 271 |
| sparsiflora............152 | rostrata..............271 |

| | |
|---|---|
| Cotton-grass............321 | Cyperus—diandrus......321 |
| Cottonwood............217 | Cypress ................328 |
| Cotula ................109 | Cypripedium.........71, 282 |
|    Australis..... ....109 |    montanum...........282 |
|    coronopifolia.........109 | Cystopteris..............334 |
| Cotyledon.......... 43, 189 |    fragilis ...............335 |
|    cæspitosa............189 | |
| Cow-parsnip........ .....176 | Dactylis ............13, 306 |
| CRASSULACEÆ............189 |    glomerata............306 |
| Cream Cups........228, 229 | Dandelion..............120 |
| Cressa..................169 | Danthonia..........13, 303 |
|    cretica ...............169 |    Californica ...........303 |
| *Croton*............79, 201 | DAPHNALES .............266 |
|    *Californicus* ..........201 | Darnel ................313 |
| CRUCIFERÆ..............221 | *Datisca*............15, 218 |
| CRUCIFLORÆ ............221 | DATISCEÆ.........• .....218 |
| CRYPTOGAMÆ............329 | Datura..............21, 154 |
| CUCURBITACEÆ..........122 |    stramonium..........154 |
| CUPULIFERÆ............268 | Daucus..............29, 176 |
| CUPRESSINEÆ............328 |    pusillus ..............176 |
| Cupressus...........73, 328 | DECANDRIA............... 39 |
|    Goveniana............328 |    MONOGYNIA............ 39 |
|    macrocarpa...........328 |    DIGYNIA ............. 41 |
| Currant................188 |    TRIGYNIA............. 41 |
| Cuscuta........18, 26, 169 |    TETRAGYNIA........... 42 |
|    salina................170 |    PENTAGYNIA. ......... 42 |
|    subinclusa............170 |    POLYGYNIA............ 43 |
| Cyanotris...........33, 287 | Deer-fern ..............333 |
|    esculenta............287 | Delphinium.........48, 233 |
| CYNAROIDEÆ............112 |    Californicum.........233 |
| Cynodon............11, 303 |    decorum.............234 |
|    dactylon.............303 |    nudicaule ...........233 |
| Cynoglossum.. .... 19, 158 |    simplex..............234 |
|    grande...............158 |    variegatum...........234 |
| CYPERACEÆ........ .....316 | Dendromecon........47, 229 |
| Cyperus........5, 9, 15, 321 |    rigidus ..............229 |
|    aristatus.............321 | Deweya.............29, 182 |

| | | | |
|---|---|---|---|
| Deweya—Hartwegi | 182 | Dirca—occidentalis | 266 |
| Kelloggii | 182 | DISCOPHORÆ | 175 |
| Deyeuxia | 302 | *Distichlis* | 307 |
| Aleutica | 302 | Dock | 274 |
| DIADELPHIA | 57 | Dodder | 169 |
| DIANDRÆ | 131 | DODECANDRIA | 43 |
| DIANDRIA | 5 | MONOGYNIA | 43 |
| DI-TRI-TETRAGYNIA | 7 | PENTAGYNIA | 44 |
| MONOGYNIA | 5 | Dodecatheon | 19, 171 |
| Dicentra | 57, 228 | Meadia | 171 |
| chrysantha | 228 | Dogbane | 129 |
| formosa | 228 | Dogwood | 185 |
| Dichondra | 168 | Douglas Spruce | 327 |
| repens | 168 | Downingia | 22, 56, 70 |
| DICHOTOMÆ | 329 | Duckweed | 325 |
| DICOTYLEDONES | 80 | Dutchman's Pipe | 268 |
| DIDYNAMIA | 49 | | |
| ANGIOSPERMIA | 51 | *Echeveria* | 43, 189 |
| GYMNOSPERMIA | 49 | *Echinocystis* | 122 |
| DIGITALEÆ | 151 | Echinospermum | 19 |
| DIŒCIA | 76 | Eel-grass | 325 |
| DIANDRIA | 76 | Elder | 124 |
| DODECANDRIA | 79 | *Eleocharis* | 9, 15, 320 |
| HEXANDRIA | 79 | Ellisia | 20, 26, 161 |
| ICOSANDRIA | 79 | chrysanthemifolia | 162 |
| OCTANDRIA | 79 | Elymus | 10, 315 |
| PENTANDRIA | 78 | arenarius | 315 |
| POLYANDRIA | 79 | condensatus | 315 |
| SYNGENESIA | 79 | Sibiricus | 315 |
| TETRANDRIA | 77 | Sitaniou | 315 |
| TRIANDRIA | 77 | Emmenanthe | 26, 159 |
| Diplacus | 52, 150 | penduliflora | 159 |
| glutinosus | 150 | ENNEANDRIA | 38 |
| DIPSACEÆ | 123 | EPIGYNÆ | 80, 282 |
| Dipsacus | 16, 123 | Epilobium | 36, 240 |
| fullonum | 123 | Franciscanum | 241 |
| Dirca | 38, 266 | minutum | 241 |

Epipactis............70, 280
   gigantea..........280
EQUISETACEÆ............335
Equisetum..............335
   Telmateia............335
   robustum.............336
Eragrostis.......... 13, 310
   poæoides.............310
Eremocarpus...... 75, 200
   setiger ..............200
ERICACEÆ...............173
ERICALES ..............172
Erigeron..........63, 68, 86
   Canadensis ........... 87
   glaucus............... 87
   Philadelphicus ....... 87
   stenophyllus.......... 86
Eriodictyon ........26, 158
   glutinosum ..........159
Eriogonum..........38, 276
   angulosum ..........276
   gracile...............277
   latifolium ...........277
   nudum ..............277
   truncatum...........277
   vimineum ...........277
   virgatum ............277
Eriophorum ......9, 15, 321
   gracile .............321
Eriophyllum ........64, 104
   cæspitosum .........105
   confertiflorum .......105
   stæchadifolium ......105
*Eritrichium*......19, 156, 157
Erodium .....24, 39, 56, 205
   botrys...............205
   cicutarium .........205

E odium-macrophyllum.205
   moschatum...........205
Eryngium...........27, 183
   petiolatum...........183
Eryngo..................183
Erysimum ........ 55, 225
   asperum ............ 225
Erythræa...........20, 130
   Douglasii. ...........131
   floribunda...........130
   Muhlenbergii........130
   trichantha..........131
   venusta.............131
Eschscholtzia....45, 47, 229
   Californica..........230
Esparto Grass .........315
Eucharidium........17, 244
   concinnum..........244
EUCYCLICÆ............200
Eunanus...........53, 151
   Douglasii............151
Euonymus.......16, 23, 214
   occidentalis .........214
EUPATORIACEÆ..........112
Euphorbia .........71, 201
   lathyris .............202
   leptocera............202
   ocellata .............201
   serpyllifolia .........201
EUPHORBIACEÆ .........200
Evax..........68, 70, 89
   caulescens........... 89
Evening Primrose.......240
Everlasting............ 90

False Indigo...........262
False Solomon's-Seal....284

| | |
|---|---|
| Fatsia....................184 | Fraxinus—Oregana......132 |
| Feather-grass..........301 | Fritillaria..........32, 290 |
| Ferula ................176 | FUMARIACEÆ............228 |
| Californica............176 | |
| Fescue Grass ..........310 | Galium..............16, 126 |
| Festuca ......5, 7, 14, 310 | Andrewsii ............127 |
| microstachys..........311 | aparine...............126 |
| myurus ..............310 | asperrimum ..........127 |
| ovina ................311 | boreale ..............126 |
| pauciflora ............311 | Californicum..........127 |
| scabrella .............311 | Nuttallii..............127 |
| FICOIDEÆ ..............190 | trifidum..............126 |
| FICOIDALES .............190 | triflorum.............127 |
| Fig-marigold ..........190 | GAMOPETALÆ............ 80 |
| Figwort................153 | Garrya............78, 185 |
| Filago..........65, 68, 89 | elliptica..............185 |
| Californica ........... 90 | Fremontii ............185 |
| FILICES................331 | Gastridium... ......12, 301 |
| FILICINEÆ...............330 | australe ..............301 |
| Fimbristylis .....9, 15, 319 | Gaultheria ..........40, 173 |
| miliacea..............319 | Shallon..............174 |
| Five-finger........233, 249 | Gayophytum.....17, 37, 241 |
| Flax.... ..........206 | diffusum.............241 |
| Fleabane....... ....... 86 | Gentian...............131 |
| Fool's Parsley ..........178 | Gentiana .....16, 20, 26, 131 |
| Foxtail......... ......298 | Oregana..............131 |
| Fragaria ...........47, 248 | GENTIANACEÆ ............130 |
| Californica ............248 | GENTIANALES............128 |
| Chilensis. ...........248 | GERANIACEÆ............204 |
| Frankenia....24, 32, 35, 219 | GERANIALES. ..........204 |
| grandiflora ..........219 | Geranium........39, 56, 205 |
| FRANKENIACEÆ..........219 | Carolinianum........205 |
| Franseria ............75, 93 | Giant Hyssop..... .....137 |
| bipinnatifida........ 93 | Gilia................21, 163 |
| Chamissonis.......... 93 | achilleæfolia..........166 |
| Fraxinus..... 6, 72, 76, 132 | androsacea...........163 |
| dipetala .............132 | capitata ..............166 |

| | |
|---|---|
| Gilia—ciliata ............164 | Grape....... ...........215 |
| cotulæfolia............165 | Gratiola...............149 |
| densifolia.............166 | ebracteata............149 |
| densiflora............163 | GRATIOLEÆ..............148 |
| dichotoma............163 | Greasewood............250 |
| inconspicua..........167 | Grindelia.........61, 64, 81 |
| intertexta............165 | cuneifolia............ 82 |
| micrantha............164 | glutinosa............. 82 |
| multicaulis...........167 | hirsutula............. 82 |
| leucocephala.........165 | robusta............... 82 |
| pusilla...............163 | Groundsel .............111 |
| squarrosa............164 | Gum-plant ............ 81 |
| tenella...............164 | Gutierrezia ..........64, 81 |
| tricolor..............167 | Californica........... 81 |
| virgata ..............166 | GUTTIFERALES...........215 |
| viscidula.............165 | Gymnogramme..........331 |
| Githopsis............22, 121 | triangularis..........331 |
| specularioides .......121 | GYMNOSPERMÆ..........326 |
| Glaux............. ..24, 171 | Gymnostichum......10, 315 |
| maritima .............171 | Californicum.........316 |
| GLUMACEÆ...............296 | GYNANDRÆ... ..........280 |
| Glyceria.............14, 308 | GYNANDRIA.............. 70 |
| pauciflora ...........308 | DIANDRIA............. 71 |
| Glycyrrhiza..........58, 262 | HEXANDRIA............ 71 |
| lepidota ..............263 | MONANDRIA............ 71 |
| Godetia .............37, 242 | |
| amœna................243 | Habenaria...........70, 281 |
| epilobioides..........243 | elegans......... .....281 |
| lepida................243 | leucostachys..........281 |
| purpurea.............243 | Hair-grass.............304 |
| quadrivulnera ........243 | HALORRHAGIDÆ.........239 |
| tenella...............243 | Hawkweed.............119 |
| Golden Aster............ 84 | Hazel ,,,,,,,,,,,,,,,,271 |
| Golden-rod............. 85 | Hedge-nettle.. .........139 |
| Gold-fern..............331 | HELENIEÆ ..............103 |
| Gooseberry ............188 | Helenium............65, 107 |
| GRAMINEÆ...............296 | puberulum... .......107 |

| | |
|---|---|
| HELIANTHEÆ............ 95 | Hetercodon—rariflorus...122 |
| Helianthella..........68, 95 | *Heteromeles* .........45, 252 |
|   Californica............. 95 | Heuchera........26, 30, 187 |
| Helianthemum......48, 217 |   micrantha............187 |
|   scoparium.............217 |   pilosissima............188 |
| Helianthella .........68, 95 | HEXANDRIA............. 31 |
|   Californica............. 95 |   HEX-POLYGYNIA........ 31 |
| Helianthus...........68, 95 |   MONOGYNIA............ 35 |
|   annuus................ 95 |   TRIGYNIA.......... ... 34 |
|   Californicus .......... 96 | Hieracium..........60, 119 |
|   exilis .. ............... 95 |   albiflorum ......... . 119 |
|   scaberrimus .......... 95 | Hierochloa..........12, 299 |
| Heliotrope ..............158 |   macrophylla..........299 |
| Heliotropium........18, 158 | Hippuris.............5, 239 |
|   curassavicum .........158 |   vulgaris............. 239 |
| HELOBIÆ ...............322 | Holcus............13, 305 |
| Hemizonia... 66, 68, 70, 98 |   lanatus...............·306 |
|   angustifolia.......... 99 | Holozonia ......66, 69, 100 |
|   corymbosa ...... ... 99 |   filipes ...............100 |
|   luzulæfolia........... 98 | Holy Grass ..........299 |
|   macradenia ... ...... . 98 | Honeysuckle ... ......125 |
|   multiglandulosa.......100 | Hop-tree .... .........208 |
|   Parryi ............... 99 | Hordeum ..... .....10, 314 |
|   pungens............. 99 |   jubatum..............314 |
|   truncata .............100 |   murinum .............314 |
| Hemlock ..............181 |   nodosum..............314 |
| Hendecandra........79, 201 | Horehound.............138 |
|   procumbens ..........201 | Horkelia.............45, 46 |
| HEPTANDRIA............. 35 | Horsetail............335 |
|   MONOGYNIA ........... 35 | Hosackia....... 58, 260, 261 |
|   POLYGYNIA............ 35 |   brachycarpa.........261 |
|   TRIGYNIA ............. 35 |   gracilis..... .........260 |
| Heracleum..........29, 176 |   parviflora............260 |
|   lanatum..............176 |   Purshiana.....  ......260 |
| Hesperocnide.......73, 273 |   stipularis............260 |
|   tenella ...............274 |   strigosa ..............260 |
| Hetercodon . .......22, 122 |   subpinnata ...........260 |

| | |
|---|---|
| Hound's-tongue ........158 | Judas Tree.......... 265 |
| Huckleberry ...........175 | JUGLANDEÆ ............209 |
| Hydrocotyle.........27, 181 | Juglans..........74, 209 |
|    prolifera .............184 |    Californica .........209 |
|    ranunculoides........181 | JULIFLORÆ..............268 |
| HYDROPHYLLEÆ .........158 | JUNCACEÆ .............293 |
| HYPERICINEÆ ...........215 | JUNCAGINEÆ .......... 323 |
| Hypericum.......46, 59, 215 | Juncus .........34, 294 |
|    anagalloides ..........216 |    Balticus..............294 |
|    concinnum ..........215 |    Breweri ..............294 |
| Hypochæris ......59, 115 |    bufonius .............295 |
|    glabra................116 |    compressus...........294 |
| HYPOGYNÆ..............128 |    effusus ..............295 |
| |    falcatus ........ 295 |
| ICOSANDRIA ............. 41 |    Kelloggii.............295 |
|    DI - TRI - TETRA - PENTA - |    Leseurii .............294 |
|    GYNIA............ 45 |    patens...............295 |
|    MONOGYNIA........... 47 |    phæocephalus.........295 |
|    POLYGYNIA........... 46 |    robustus ............294 |
| Indian Hemp......129, 130 |    tenuis ........... 295 |
| INULEÆ ................ 88 |    xiphoides............295 |
| IRIDACEÆ.............282 | June-Berry...........252 |
| Iris........ .......8, 283 | Juniper........... 328 |
|    Douglasiana..  .....283 | Juniperus........78, 328 |
|    longipetala ..........283 |    Californica ..........329 |
|    macrosiphon.........283 | Jussiæa ........36, 40, 240 |
| ISOCARPICÆ .............170 |    repens .......... 240 |
| ISOËTEÆ ..........329 | |
| Isoëtes .................329 | Knot-grass............275 |
| Isolepis ..............319 | Kœleria...........13, 306 |
|    carinata...............319 |    cristata..............307 |
|    riparia ...............319 | Krynitzkia ...... 19, 156 |
| Iva..........66, 70, 76, 92 |    ambigua. ....... 156 |
|    axillaris ............. 92 |    Californica...........157 |
| |    Chorisiana............ 57 |
| Jacob's-ladder .........163 |    leiocarpa ........ 156 |
| *Jaumea* ...............103 |    muriculata..........156 |

| | |
|---|---|
| Krynitzkia—oxycarya....157 | Lemna—gibba....... . 326 |
| Torreyana..........156 | minor...............326 |
| | trisulca . ............325 |
| LABIATÆ ...............132 | Valdiviana . .........325 |
| Ladies'-Tresses ........281 | LEMNACEÆ....... .....325 |
| Lady's-Mantle ..........250 | LENTIBULARIA ........ 170 |
| Lady's-slipper ..........282 | Lepidium..........6, 54, 222 |
| Lagophylla.......67, 69, 100 | latipes................223 |
| congesta...............101 | nitidum..............223 |
| ramosissima....... ...101 | Lepigonum...............194 |
| LAMIALES. ............ 132 | Leptosyne........ ....66, 96 |
| Larkspur...............233 | Stillmani............. 96 |
| Lastarriæa ..........15, 278 | Lepturus............10, 312 |
| Chilense..............278 | incurvatus........... 313 |
| Lasthenia...........64, 105 | Lessingia........61, 63, 83 |
| Californica............106 | Germanorum.......... 83 |
| glaberrima............106 | leptoclada ............ 83 |
| glabrata..............106 | ramulosa............. 83 |
| Lathyrus...........58, 265 | Libocedrus.............. 73 |
| littoralis............. 265 | LIGULIFLORÆ............115 |
| palustris .............265 | Ligusticum. ..........178 |
| vestitus ..............265 | apiifolium............178 |
| LAURINEÆ ..............230 | Lilæa..............71, 324 |
| Lavatera .... .57, 202 | subulata..... ........325 |
| assurgentiflora........230 | LILIACEÆ..............285 |
| Layia.. ............67, 101 | Lilium............33, 289 |
| calliglossa.............102 | pardalinum...........290 |
| chrysanthemoides.....102 | Lily.. ................289 |
| carnosa ...............101 | LIMNANTHEÆ............207 |
| elegans...............101 | Limnanthes........40, 207 |
| gaillardioides... .....102 | alba.................208 |
| heterotricha..........101 | Douglasii.............208 |
| hieracioides ..........102 | Limosella...........51, 148 |
| platyglossa.,.........102 | aquatica.............148 |
| Leatherwood...........266 | Linaria............53, 153 |
| LEGUMINOSÆ............253 | Canadensis...........153 |
| Lemna..............6, 325 | LINEÆ.. ..... .........206 |

| | |
|---|---|
| Linum .........30, 31, 206 | Lupinus—nanus ........255 |
| Breweri ............207 | polyphyllus..........254 |
| Californicum.........207 | rivularis............254 |
| congestum...........207 | trifidus ............256 |
| perenne .............206 | truncatus ...........256 |
| spergulinum........207 | Luzula............34, 296 |
| Liquorice..........362 | comosa.............296 |
| Lip-fern ............331 | LYCOPODIACEÆ .........329 |
| Lippia ...........53, 140 | Lycopus.............7, 133 |
| nodiflora...........140 | lucidus.............133 |
| LOASACEÆ .............217 | LYTHRARIÆ............245 |
| Loco-weed..............263 | Lythrum........32, 44, 245 |
| Lolium..............10, 313 | alatum..............245 |
| perenne .............313 | |
| temulentum .........313 | Madia.........62, 66, 70, 97 |
| *Lomaria*................333 | dissitiflora............97 |
| Lonicera ...........22, 125 | elegans..............97 |
| involucrata..........125 | filipes...............98 |
| hispidula...........125 | Nuttallii.............97 |
| Loose-strife............245 | radiata ..............97 |
| Lophanthus.........50, 137 | sativa................97 |
| urticifolius..........137 | Madroña...............174 |
| Lophochlæna........14, 308 | Maianthemum.......17, 284 |
| Californica...........308 | bifolium.............284 |
| LORANTHACEÆ...........266 | Maiden's-Hair .........333 |
| Lovage.................178 | Malacothrix........59, 118 |
| Lupine.................253 | Californica...........118 |
| Lupinus....... 56, 57, 253 | Clevelandi...........118 |
| affinis................255 | obtusa...............118 |
| albicaulis.............255 | Mallow................203 |
| arboreus .............254 | Malva.............57, 203 |
| Chamissonis ..........254 | borealis .............203 |
| densiflorus...........256 | MALVACEÆ .  .........202 |
| Douglasii. ...........254 | MALVALES .............202 |
| littoralis.............255 | Malvastrum ...........204 |
| micranthus...........256 | Thurberi.............204 |
| microcarpus .........256 | Manzanita..............174 |

| | |
|---|---|
| Maple...................211 | Mentha—viridis .......133 |
| Mare's-tail..............239 | Menyanthes........20, 131 |
| Mariposa...............290 | trifoliata..............131 |
| Marrubium......... 49, 138 | Mentzelia...... 46, 59, 218 |
| vulgar..............139 | lævicaulis............218 |
| Marsh-fleabane.......... 88 | Lindleyi..............218 |
| Marsilæa................330 | Mesembryanthemum..46, 47 |
| vestita................330 | 190. |
| Matricaria..........62, 108 | æquilaterale..........190 |
| discoidea.............108 | MICRANTHÆ.............296 |
| occidentalis...........108 | Microcala...........16 130 |
| Mayweed................107 | quadrangularis........130 |
| Meadow-rue............ 237 | Micromeria...... 50, 135 |
| Meconopsis..........47, 230 | Douglasii.............135 |
| heterophylla..........230 | Micropus............65, 88 |
| Medicago ....... 58, 259 | Californicus...........88 |
| denticulata............259 | Microseris......59, 116, 117 |
| lupulina..............259 | acuminata............117 |
| sativa................259 | aphantocarpha........118 |
| Medick.................259 | attenuata............117 |
| Megarrhiza.......73, 122 | Bigelovii.............118 |
| *Californica*............123 | Douglasii............117 |
| *fabacea*...............123 | elegans..............118 |
| *marah*...............123 | Milkweed............ 128 |
| MELANTHACEÆ......... 29 | Milkwort..............171 |
| Melica............13, 307 | Mimetanthe....... 52, 149 |
| bromoides............307 | pilosa...............149 |
| bulbosa..............307 | Mimulus.........52, 54, 149 |
| Harfordii............. 307 | cardinalis............150 |
| imperfecta............307 | inconspicuus.........149 |
| Melic-grass............307 | luteus...............150 |
| Melilot ................258 | moschatus............150 |
| Melilotus...........58, 258 | nasutus..............149 |
| parviflora............259 | Miner's-lettuce ......195 |
| Mentha.............50, 133 | Mint.................133 |
| Canadensis...........133 | Mistletoe.............267 |
| piperita..............133 | Mistletoe (pine)........267 |

| | |
|---|---|
| Monadelphia............ 56 | Muilla...............32, 289 |
|    decandria ...  ...... 56 |    maritima.............289 |
|    octandria............ 56 | Mustard. .............225 |
|    pentandria .......... 56 | Myosurus........31, 49, 236 |
|    polyandria ......... 57 |    minimus ... ........236 |
|    triandria....... .... 56 | Myricaceæ.............272 |
| Monandria............. 5 | Myrica..............76, 272 |
|    di-tri-tetragynia .... 5 |    Californica...........272 |
|    monogynia............ 5 | Myriophyllum.......74, 239 |
| Monardella....... ....50, 134 |    spicatum........... .. 239 |
|    Breweri .............134 | Myrtales ..............238 |
|    Douglasii............134 | |
|    lanceolata ............134 | Naiadaceæ.............324 |
|    undulata.............134 | Naias (Najas)........... 71 |
|    villosa...............134 | Nardosmia .........69, 110 |
| Monkey-flower.. ... 149, 150 |    palmata..... .......111 |
| Monochlamydeæ........267 | Nasturtium......  ....55, 223 |
| Monocotyledones.......280 |    officinale..............223 |
| Monœcia ........ ...... 71 | Negundo..18, 25, 77, 78, 211 |
|    diandria............. 72 |    Californicum..........212 |
|    monandria............ 71 | Neillia..............45, 247 |
|    pent—polyandria..... 74 |    opulifolia.............247 |
|    tetrandria........... 73 | Nemophila...........20, 26 |
|    triandria....... .... 72 | Nettle..................273 |
| Monogamia............. 70 | Nicotiana ....... ...21, 154 |
| Monolopia .. ......64, 105 |    Bigelovii.............155 |
|    gracilens.......... ...105 | Nightshade.............153 |
|    major........ ......105 | Nine-bark .............247 |
| Monterey-pine .........327 | Nuphar ...............238 |
| Montia .... ........8, 196 |    polysepalum..........238 |
|    fontana..............197 | Nutmeg Pine............326 |
| Morning-glory...........168 | Nuttallia .......44, 45, 246 |
| Mountain Mahogany....250 |    cerasiformis ..........246 |
| Mountain Mint.........133 | Nyctagineæ............199 |
| Mouse-ear Chickweed ...192 | Nymphæaceæ..........238 |
| Mousetail .............236 | |
| Mudwort..... ........148 | Oak................ ...269 |

| | |
|---|---|
| OCTANDRIA............. 36 | Panicum............10, 296 |
| DI-TRIGYNIA........... 38 | PAPAVERACEÆ..........228 |
| MONOGYNIA............. 36 | PAPILIONACEÆ ..........253 |
| TETRAGYNIA,.......... 38 | PARIETALES............217 |
| Œnanthe.........28, 178 | PARONYCHIÆ ............194 |
| Californica.......... 178 | Pearly-Everlasting...... 90 |
| Œnothera..........37, 241 | Pearlwort .............193 |
| biennis.............241 | Pectocarya........ 19, 157 |
| Californica..........242 | penicillata. ..........158 |
| Cheiranthifolia.......242 | Pedicularis .........51, 143 |
| dentata.............242 | densiflora ...........143 |
| micrantha ..........242 | *Pellœa*................332 |
| ovata ..............242 | Pentacœna....14, 24, 27, 194 |
| strigulosa .........242 | ramosissima..........195 |
| OLEACEÆ..............131 | Pentachæta...... 61, 63, 82 |
| ONAGRACEÆ............239 | alsinoides ........... 83 |
| Onion................286 | bellidiflora........... 82 |
| OPHIOGLOSSEÆ .........335 | exilis ............... 83 |
| Orache...............198 | PENTANDRIA ........... 18 |
| Orchard grass.........306 | DIGYNIA.............. 25 |
| ORCHIDACEÆ...........280 | MONOGYNIA ....... 18 |
| Oreodaphne.........38, 231 | PENTAGYNIA.......... 31 |
| Californica...........231 | POLYGYNIA........... 31 |
| OROBANCHACEÆ........142 | TRIGYNIA............. 31 |
| Osmorrhiza.........29, 179 | Pentstemon......21, 53, 151 |
| brachypoda..........179 | azureus .............151 |
| nuda................179 | centranthifolius......151 |
| OXALIDEÆ .............206 | Peony ................232 |
| Oxalis..........43, 56, 206 | Peppergrass .......... 222 |
| corniculata..........206 | PRIGYNÆ .............238 |
| Oregana.............206 | PERSONALES...........142 |
| | *Petasites* ..............110 |
| Pæonia..........49, 232 | Peucedanum ........29, 177 |
| Brownii.............233 | caruifolium...........177 |
| Painted-Cup..........147 | dasycarpum .........177 |
| Panax............27, 184 | leiocarpum ..........177 |
| horridum............184 | triternatum..........177 |

Peucedanum — utriculatum..............177
Phacelia............26, 160
　ciliata..................161
　circinata............160
　divaricata............161
　distans................10
　hispida..............160
　malvæfolia..........160
　tanacetifolia..........161
Phalaris..........11, 298
　amethystina..........299
　arundinacea..........299
　Canariensis..........298
　intermedia............299
PHANEROGAMÆ........80
PHILADELPHEÆ..........189
Phleum..........11, 297
　alpinum..............297
　pratense..............297
Phoradendron...73, 77, 267
　flavescens............267
Photinia............45, 252
　arbutifolia............252
Phragmites..........12, 306
　communis............306
Pickeringia..........39, 253
　montana..............253
Pigweed..............197
Pilularia..............330
　Americana............330
Pimpernel..............172
Pimpinella..........28, 180
　apiodora..............181
Pin-clover..............205
Pine..................326
Pinus............72, 326

Pinus—insignis........327
　muricata.............327
　Sabiniana.............327
　tuberculata...........327
PIPERALES..............279
Plagiobothrys......19, 157
　canescens............157
　nothofulvus..........157
PLANTAGINEÆ..........140
Plantago..........16, 141
　lanceolata............141
　major................141
　maritima.............141
　Patagonica...........141
Plantain..............141
PLATANEÆ..............272
Platanus..........76, 273
　racemosa.............273
Platystemon........48, 228
　Californicus..........229
Platystigma........48, 229
　Californicum..........229
　lineare..............229
Plectritis............8, 124
　congesta.............124
　macrocera............124
Pluchea............65, 88
　camphorata............88
Plum..................246
PLUMBAGINEÆ..........170
Poa..............13, 309
　annua...............309
　Californica..........309
　distans..............309
　Douglasii............310
　pratensis............309
　scabrella............309

| | |
|---|---|
| Poa—tenuifolia..........309 | Pondweed..............323 |
| trivialis ..............309 | Poplar...................217 |
| Pogogyne.......... 50, 135 | Populus.............79, 217 |
| Douglasii.............135 | trichocarpa...........217. |
| parviflora............135 | POTAMEÆ...............323 |
| serpylloides...........135 | Potamogeton .......18, 323 |
| POLEMONIACEÆ ........162 | lucens................324 |
| POLEMONIALES.. ........153 | natans ...............324 |
| Polemonium........21, 163 | pauciflorus ...........324 |
| carneum..............163 | pectinatus............324 |
| POLYADELPHIA.......... 59 | Potentilla ..........46, 249 |
| POLYANDRIA............ 47 | anserina..............249 |
| DI - TRI - TETRA - PENTA - | glandulosa...........249 |
| POLYGYNIA .......... 48 | Portulaca, 35, 36, 42, 43, 45 |
| MONOGYNIA........... 47 | 195. |
| POLYCARPICÆ.......230, 322 | oleracea ..............195 |
| Polygala. . .....56, 57, 210 | PORTULACACEÆ..........195 |
| cucullata..............210 | Prickly Poppy..........230 |
| POLYGALACEÆ............210 | PRIMULACEÆ............171 |
| POLYGONACEÆ ..........274 | PRIMULALES ............170 |
| Polygonum......24, 38, 275 | Prosartes............33, 284 |
| aviculare ............275 | Hookeri..............285 |
| Californicum..........275 | Menziesii.............285 |
| convolvulus..........276 | Prunella ...........50, 138 |
| Muhlenberghii.......276 | vulgaris..............138 |
| nodosum.............276 | Prunus.................246 |
| paronychia...........275 | demissa ..............246 |
| POLYPETALÆ.............175 | emarginata...........246 |
| POLYPODIACEÆ .........331 | ilicifolia...............246 |
| Polypodium............ 332 | Pseudotsuga............327 |
| Scouleri..............332 | Douglasii.............327 |
| vulgaris..............332 | Psilocarphus .....67, 70, 89 |
| Polypody...............332 | Oreganus............. 89 |
| Polypogon ........11, 300 | tenellus.............. 89 |
| littoralis..............300 | Psoralea.........57, 58, 261 |
| Monspeliensis ........300 | macrostachya.........262 |
| POMACEÆ................252 | orbicularis...........261 |

Psoralea—physodes.....262
   strobilina. ...........262
Ptelea, 17, 24, 73, 74, 77, 78
   208.
   angustifolia...........208
Pteris..................332
   aquilina..............333
Pterostegia..........34, 278
   drymarioides..........278
Pugiopappus.........66, 96
   calliopsideus.......... 96
Purslane................195
Pycnanthemum......50, 133
   Californicum..........134
Pyrola..............39, 173
   aphylla...............173
PYROLACEÆ.. ...........172

Quaking-grass ..........310
Quercus.............76, 269
   agrifolia..............270
   chrysolepis........... 269
   densiflora ............270
   Douglasii.............269
   dumosa ..............269
   Kelloggii..............270
   lobata................269
   Wislizenii ............270
Quillwort. ............329
Quinine-Tree............185

Radish ................221
Rafinesquia..........59, 115
   Californica...........115
Ragweed .............. 92
RANALES............ ...237
RANUNCULACEÆ..........232

Ranunculus.........49, 235
   aquatilis..............236
   Bloomeri........ .....235
   Californicus ..........235
   flammula.............236
   hebecarpus ..........235
   Lobbii.... ..........236
   maximus.............235
   muricatus............235
   pusillus..............236
Raphanus...........54, 221
   sativus...............222
Raspberry..............247
Rattleweed.............263
Redwood....  ........328
Reed...................306
RHAMNACEÆ ...........212
Rhamnus........23, 78, 212
   Californica ...........212
   Crocea ..............212
   Purshiana............212
RHINANTHEÆ . .........143
RHIZANTHEÆ............267
RHIZOCARPÆ...........330
Rhododendron ......40, 173
   occidentale ..........173
RHODORACEÆ............173
Rhus......30, 74, 78, 82, 209
   aromatica............209
   diversiloba...........209
Ribes.. .. ......24, 30, 188
   aureum...............188
   divaricatum ..........188
   Menziesii.............188
   sanguineum .. .......188
Rigiopappus.........64, 106
   leptocladus—see *Errata*.

| | |
|---|---|
| Rock-brake............332 | Salix—lasiolepis........216 |
| Rock-cress.............226 | longifolia ...........216 |
| Rock-rose .............217 | Salvia...............7, 136 |
| Romanzoffia.........20, 159 | carduacea ...........136 |
| Sitchensis ...........159 | Columbariæ .........137 |
| Rosa................46, 251 | SALVINIACEÆ ..........330 |
| Californica ..........251 | Sambucus ..........30, 124 |
| gymnocarpa .........251 | glauca..............124 |
| Rose...................251 | Samolus.........22, 41, 172 |
| ROSACEÆ................251 | Valerandi ...........172 |
| ROSIFLORÆ.............245 | Samphire.............199 |
| RUBIACEÆ...............126 | Sand-spurrey ...........194 |
| RUBIALES...............133 | Sandwort..............193 |
| Rubus..............46, 247 | Sanicle ..............182 |
| leucodermis..........248 | Sanicula ............27, 182 |
| Nutkanus ...........247 | Sanguisorba ..........18 |
| spectabilis ..........247 | SANGUISORBACEÆ ........249 |
| ursinus .............248 | SAPINDACEÆ ............210 |
| Rumex .........34, 79, 274 | SAPINDALES ............210 |
| acetosella ...........275 | SAURUREÆ ............279 |
| conglomeratus ......275 | Saxifraga.......... 41, 186 |
| crispus..............274 | Virginica...........186 |
| obtusifolius ........275 | SAXIFRAGACEÆ..........186 |
| salicifolius.........274 | Saxifrage .............186 |
| RUTACEÆ...............208 | Scirpus........6, 15, 69, 320 |
| | acicularis............320 |
| Sage ..................136 | lacustris........ ..320 |
| Sage-brush.............109 | maritimus ...........320 |
| Sagina..........31, 42, 193 | Olneyi .............320 |
| occidentalis .........193 | palustris ............320 |
| SALICINEÆ ..............216 | Tatora ..............320 |
| Salicornia...........5, 199 | sylvaticus...........320 |
| ambigua.............199 | Scoliopus ...........8, 292 |
| Salix............76, 77, 216 | Bigelovii ............292 |
| Coulteri..............217 | Scorzonella............116 |
| lævigata .......... ..216 | paludosa ............116 |
| lasiandra............216 | procera .............116 |

| | |
|---|---|
| Scorzonella—sylvatica .. 116 | Shepherd's-purse . ..... 223 |
| Scouring-rush .......... 335 | Shield-fern............. 334 |
| Scrophularia ........ 53, 153 | Shooting-star.......... 171 |
|    Californica ........... 153 | Sida..... ..........57, 204 |
| SCROPHULARIÆ ........ 143 |    hederacea ........... 204 |
| Scutellaria.......... 50, 138 | Sidalcea............. 57, 203 |
|    Californica ........... 138 |    diploscypha ......... 203 |
|    tuberosa.............. 138 |    humilis .............. 203 |
| Sea-blite............... 199 |    malachroides ........ 203 |
| Sea-lavender .......... 171 |    *malva:flora* .......... 203 |
| Sea-Rocket ............ 224 | Silene.............. 41, 191 |
| Sedge ................. 316 |    antirrhina........... 191 |
| Sedum.............. 43, 189 |    Californica .......... 191 |
|    spathulifolium ....... 189 |    Gallica ........ ... 191 |
|    stenopetalum ........ 189 |    laciniata............ 191 |
| Selaginella............ 329 |    verecunda........... 192 |
|    rupestris ............ 329 | SILENEÆ................ 191 |
| SELAGINELLEÆ.......... 329 | Silybum............... 114 |
| Self-heal ............. 138 |    Marianum........... 114 |
| Selinum............ 28, 178 | Sisymbrium......... 56, 224 |
|    Pacificum ........... 178 |    acutangulum ....... 224 |
| Senebiera........... 54, 222 |    officinale..... ........ 224 |
|    didyma.............. 222 |    reflexum ............ 224 |
| Senecio.... ..... 61, 65, 111 | Sisyrinchium........ 56, 282 |
|    aronicoides.......... 111 |    bellum .............. 283 |
|    Douglasii ........... 112 |    Californicum ........ 283 |
|    eurycephalus........ 111 | Sium .............. 28, 179 |
|    hydrophilus ......... 111 |    cicutæfolium........ 179 |
|    vulgaris.......... ... 112 | Skullcap .............. 138 |
| SENECIONIDÆ ........... 110 | SMILACEÆ .............. 284 |
| Sequoia............. 74, 328 | Smilacina ........... 34, 284 |
|    gigantea............. 328 |    amplexicaulis........ 284 |
|    sempervirens......... 328 |    sessilifolia........... 284 |
| SERPENTARIÆ.......... 267 | Snapdragon ........... 152 |
| Service-berry........... 252 | Sneezeweed ........... 107 |
| Setaria ................ 10 | Snowberry............. 125 |
| Sheep-sorrel........ 274, 275 | Soap-root........... .. 286 |

| | |
|---|---|
| SOLANACEÆ ............ 153 | Spiranthes—porrifolia ... 281 |
| Solanum ........... 22, 153 | Romanzoffiana ...... 281 |
| nigrum ............... 154 | Spleenwort ............ 334 |
| umbelliferum ........ 154 | Spotted Thistle ......... 114 |
| Solidago ............ 63, 85 | Spurge ................ 201 |
| Californica .......... 85 | Squirrel-tail ........... 314 |
| occidentalis .......... 85 | St. John's-wort ........ 215 |
| sempervirens ......... 85 | Stachys ........ ...... 50, 139 |
| Soliva ............. 69, 109 | ajugoides ............ 139 |
| sessilis .............. 109 | albens ............... 139 |
| Sonchus ........... 60, 120 | bullata ............... 140 |
| asper ................ 120 | Chamissonis ......... 140 |
| oleraceus ............ 120 | pycnantha ........... 139 |
| Sow-thistle ............ 120 | Star-flower .......... 171 |
| SPADICIFLORÆ .......... 321 | Star-thistle ........... 114 |
| Sparganium ......... 72, 322 | Statice ............ 31, 171 |
| eurycarpum ......... 322 | limonium ............ 171 |
| Spartina ............ 11, 202 | Stellaria .......... 42, 192 |
| stricta .............. 203 | media ............... 192 |
| Specularia .......... 22, 121 | nitens ............... 193 |
| biflora ............... 121 | Stephanomeria .. .. 59, 115 |
| Speedwell ............. 148 | virgata .............. 115 |
| Spergula ....... 31, 43, 194 | Stipa ............. 12, 301 |
| arvensis ............. 194 | eminens ............. 302 |
| Spergularia ........ 42, 194 | setigera ............. 302 |
| macrotheca .......... 194 | viridula ............. 302 |
| media ............... 194 | STIPULATÆ ............. 335 |
| Sphacele ........... 50, 136 | Stone-crop ........... 189 |
| calycina ............. 136 | Stork's-bill ........... 205 |
| *Sphaeralcea* ........ 57, 204 | Strawberry ............ 248 |
| Spice-bush ............ 252 | Streptanthus ....... 55, 226 |
| Spikenard ............. 184 | glandulosus ......... 226 |
| Spindle-tree .......... 214 | niger ................ 226 |
| Spiræa ......... 45, 46, 247 | peramœnus .......... 226 |
| discolor ............. 247 | Stylocline ....... 65, 67, 70 |
| SPIRÆEÆ ............. 247 | Suæda ............ 25, 199 |
| Spiranthes ......... 71, 281 | Californica .......... 199 |

17

| | | | |
|---|---|---|---|
| Sunflower | 95 | TETRADYNAMIA | 54 |
| Sweet Cicely | 179 | SILICULOSA | 54 |
| Sweet Vernal-grass | 299 | SILIQUOSA | 54 |
| Sycamore | 273 | TETRANDRIA | 16 |
| Symphoricarpus | 22, 125 | DIGYNIA | 16 |
|    mollis | 125 | MONOGYNIA | 16 |
|    racemosus | 125 | Thalictrum | 49, 79, 237 |
| SYNANDRÆ | 80 |    Fendleri | 237 |
| SYNGENESIA | 59 | Thermopsis | 39, 253 |
|    ÆQUALIS | 59 |    Californica | 253 |
|    FRUSTRANEA | 68 | Thistle | 112 |
|    NECESSARIA | 68 | Thorn-apple | 154 |
|    SUPERFLUA | 62 | Thrift | 170 |
| Synthyris | 7, 148 | THYMELACEÆ | 266 |
|    rotundifolia | 148 | Thysanocarpus | 54, 222 |
| Syrmatium | 261 |    curvipes | 222 |
|    cytisoides | 261 |    laciniatus | 222 |
|    glabrum | 261 |    pusillus | 222 |
|    tomentosum | 261 | Tiarella | 41, 187 |
| | |    unifoliata | 187 |
| Tanacetum | 61, 109 | Tillæa | 14, 18, 31, 190 |
|    camphoratum | 109 |    augustifolia | 190 |
| Tansy | 109 |    minima | 190 |
| Taraxacum | 60, 120 | Timothy | 297 |
|    officinale | 120 | Toad-Flax | 153 |
| Tar-weed | 97, 98 | Tobacco | 154 |
| Tassel-tree | 185 | Torreya | 79, 326 |
| TAXINEÆ | 326 |    Californica | 326 |
| Teasel | 123 | Toyon | 252 |
| Tellima | 41, 186 | Tree-Mallow | 202 |
|    affinis | 187 | TRIANDRIA | 8 |
|    Bolanderi | 187 |    DIGYNIA | 9 |
|    grandiflora | 187 |    MONOGYNIA | 8 |
|    heterophylla | 187 |    TRIGYNIA | 14 |
| TEREBINTHACEÆ | 208 | Tricerastes | 15, 218 |
| TEREBINTHALES | 208, 209 |    glomerata | 218 |
| TETRACYCLICÆ | 80 | Trichostema | 51, 132 |

Trichostema—lanceolatum .................132
TRICOCCÆ ...............200
Trientalis........32, 35, 171
   latifolia ...............172
Trifolium.........58, 256
   barbigerum...........258
   bifidum .............257
   ciliatum..............257
   depauperatum .......258
   fucatum ...........258
   gracilentum..........257
   involucratum .......257
   Macræi...............257
   microcephalum ......258
   microdon ...........258
   pauciflorum .........258
   repens................257
   tridentatum .........257
Triglochin ......34, 35, 323
   maritimum..........323
Trillium........32, 33, 285
   ovatum..............285
   sessile ..............285
Trisetum...............304
   barbatum............304
   canescens............304
Triteleia...........32, 288
   capitata .............288
   ixioides .............289
   lactea ...............289
   laxa..................289
   peduncularis........289
Triticum............10, 314
   repens...............314
Tropidocarpum......55, 223
   gracile...............223

Troximon..........60, 119
   apargioides..........119
   grandiflorum.........119
   heterophyllum........120
   humile...............119
   laciniatum...........119
TUBULIFLORÆ........... 81
*Turritis glabra*.........227
Twin-berry............. 125
Typha ............72, 322
   latifolia ............. 322
TYPHACEÆ..............321

UMBELLALES...........175
UMBELLIFERÆ..........175
*Umbellularia*............231
URTICALES............272
Urtica ........73, 77, 273
   holosericea..........273
   Lyallii...............273
   urens................273
Utricularia...........7, 170
   vulgaris..............170

VACCINEÆ.............175
Vaccinium .........37, 175
   ovatum..............175
VALERIANACEÆ.........124
Vancouveria........31, 231
   hexandra............231
Venus Looking-glass....121
VERBASCEÆ.............153
Verbena............53, 140
   officinalis............140
VERBENACEÆ..........140
Veronica............6, 148
   Americana..........148

| | |
|---|---|
| Veronica—peregrina ....148 | Wild-Ginger.............268 |
| VERONICÆ .......... .. 147 | Wild-Pea..............265 |
| Vetch ...............264 | Wild-Rye..............315 |
| Vicia...... ........58. 264 | Willow ...... ......216 |
|    Americana ............264 | Willow-herb..........240 |
|    exigua............. 264 | Winter cress ..........224 |
|    gigantea.............264 | Wintergreen ... ...173 |
|    sativa...............264 | Wood-sorrel ...........206 |
| Viola ..... ......23, 220 | Wood-rush..............296 |
|    aurea ........ ...... 220 | Woodwardia...........333 |
|    canina................220 |    radicans.............334 |
|    lobata... ............221 | Wormwood. ........ ...109 |
|    ocellata ........ ... 220 | *Wulfenia*..............148 |
|    odorata... ...........220 | Wyethia ............66, 94 |
|    pedunculata ......220 |    angustifolia........... 95 |
|    sarmentosa ...........221 |    glabra ............. 94 |
| VIOLACEÆ................219 |    helenioides........... 94 |
| Virgin's Bower..........237 | |
| Vitis ............23, 78, 215 | Xanthium.............. 93 |
|    Californica . .........215 |    spinosum............. 93 |
| |    strumarium .... ..... 93 |
| Wake-robin .............285 | Xerophyllum ...........293 |
| Waldmeister............127 |    tenax ................293 |
| Wall-flower............226 | |
| Walnut ................209 | Yarrow .... ..... .... 108 |
| Water-cress.............323 | Yedra .. ............209 |
| Water-fern. ............331 | Yellow-dock ...........275 |
| Water-hemlock..........180 | Yerba buena...........135 |
| Water-horehound ......133 | Yerba santa............158 |
| Water-lily...... ... .. 238 | Yerba mansa . .........279 |
| Water-milfoil ..........239 | |
| Water-parsnip .........179 | Zauschneria ............240 |
| Water-pennywort ......184 |    Californica...........240 |
| Water-plantain.........323 | Zostera.. .... ......71, 325 |
| Wax-Myrtle ............272 |    marina..............325 |
| Wheat .......... ....314 | Zygadene.............292 |
| Whipples.. ... 30, 42, 189 | Zygadenus..........34, 292 |
|    modesta..............189 |    Fremontii ............292 |
| |    venenosus. ..........292 |

# APPENDIX.

NOTE.—In this work, in order to conform with most American treatises on botany, the word *loculus* is replaced by the less exact word *cell*.

⊙. Annual.
⊙⊙. Biennial.
♃. Perennial.
♄. Shrub or tree.
♂. Male.
♀. Female.
☿. Hermaphrodite, *i. e.*, both sexes in the same flower.
∞. Many, or an indefinite number.

*Acaulescent.* Stemless.
*Accrescent.* Increasing with age.
*Acerose.* Needle-shaped.
*Aculeate.* Having sharp points or prickles.
*Acuminate.* Tapering to a point.
*Æqualis.* All flowers hermaphrodite.
*Acute.* Sharp-pointed.
*Adnate.* United by the surface.
*Akene.* A dry, hard, indehiscent, 1-celled, 1-seeded fruit.
*Alveolate.* Pitted like a thimble
*Ament, Amentum.* A unisexual spike with scaly bracts.
*Amplexicaul.* Clasping the stem.
*Anceps, Ancipital.* Two-edged.
*Andrœcium.* Stamens and corolla.
*Androgynous.* Having both male and female flowers.
*Angiosperm, Angiospermæ.* Bearing seeds in a closed pericarp.
*Anisocarpæ.* Number of Carpels less than lobes of Corolla.
*Annular.* Ring-like.
*Anther.* The part of the stamen bearing pollen.
*Aphyllous.* Without leaves.

*Apetalous.* Without petals.
*Apocarpous.* Carpels (Carpidia) distinct.
*Apophyses.* Enlargements, structurally united.
*Appressed.* Pressed close.
*Arachnoid.* Cobweb-like.
*Arillus.* An expanded appendage attached to the seed.
*Aristate.* Having an awn.
*Articulate.* Jointed.
*Ascending.* Rising, but not erect.
*Aspergilliform.* Brush-like.
*Auriculate.* Having an ear-like lobe at the base.
*Awn.* A bristle-like appendage.
*Axil.* The angle formed by a leaf or branch with the stem.

*Baccate.* Berry-like.
*Barbellate.* Having minute awns with reflexed points.
*Basifixed.* Attached by the lower end.
*Bifid.* Two-cleft.
*Bipinnate.* Twice pinnate.
*Bract.* A leaf, or modification of one, subtending a flower or cluster.
*Bracteolate.* Having secondary bracts upon the pedicel.

*Caducous.* Falling very early.
*Cæspitose, Cespitose.* Growing in tufts.
*Calamus.* A stuffed scape.
*Calcarate.* Spurred.
*Calycine.* Calyx-like.
*Calyculate.* Calyx-like.
*Campanulate.* Bell-shaped.
*Canaliculate.* Fluted.
*Canescent.* Hoary with a gray pubescence.
*Capillary.* Very slender and hair-like.
*Capitate.* Collected into a head.
*Capitulate.* Diminutive of capitate.
*Carina.* Keel.
*Carpidia.* A simple pistil, or one of the several parts of a compound pistil.

*Carpophore.* A prolongation of the axis between the carpels.

*Caryopsis.* Seed inseparably united to the wall of the ovary.

*Caudate.* Tailed.

*Caulescent.* Having a manifest stem.

*Cauline.* Belonging to the stem.

*Centrospermæ.* Seeds affixed to the centre of a 1-celled ovary.

*Chlorophyll.* The green coloring matter of plants.

*-chotomous.* Forked.

*Ciliate.* Fringed with hairs.

*Circinate.* Coiled from the tip.

*Circumscissile.* Opening by a transverse circular line.

*Cirrhate.* Tendril-bearing.

*Clavate.* Club-shaped.

*Cleistogamic.* Having fertile but undeveloped flowers.

*Coccus.* The indehiscent segment of a compound ovary.

*Cœlospermæ.* Internal surface of Endosperm concave.

*Commissure.* The surface by which two carpels adhere.

*Complicate.* Folded together.

*Conduplicate.* Doubled together lengthwise.

*Connate.* United in one.

*Connective.* The part of the filament which connects the anther cells.

*Contorted.* Twisted.

*Convolute.* Rolled together from one edge.

*Cordate.* Heart-shaped.

*Coriaceous.* Leathery.

*Corniform.* Horn-shaped.

*Corolline.* Like a corolla.

*Corona.* Crown.

*Corymb.* A depressed raceme.

*Costa.* A rib.

*Cotyledon.* The seed leaf or lobe of the embryo.

*Cremocarp.* Fruit of Umbelliferæ.

*Crenate.* Having rounded teeth.

*Crenulate.* Finely crenate.
*Cryptogamæ.* Fructifying without stamens and pistils.
*Cucullate.* Like a cowl or hood.
*Cuneate.* Wedge-shaped enlarging upward.
*Cuneiform.* Same as cuneate.
*Cupula.* A cup-shaped involucre inclosing a nut as in the acorn.
*Cusp.* A sharp rigid point.
*Cyathiform.* Cup-shaped.
*Cyme.* A broad and flattish inflorescence, flowering from the centre outward.

*Decandria* Ten stamens.
*Declined.* Bent or curved downward.
*Decompound.* Repeatedly compound or divided.
*Decumbent.* Reclining at base, the summit ascending.
*Decurrent.* Running down the stem.
*Decussate.* In pairs or threes alternating at right angles.
*Deflexed.* Bent or turned down abruptly.
*Dehiscent.* Opening regularly by valves, slits, etc.
*Deltoid.* Broadly triangular like the Greek letter Delta.
*Dentate.* Toothed.
*Denticulate.* Minutely toothed.
*Diadelphia.* Stamens in two sets or clusters.
*Diandria* Two stamens.
*Dichotomous.* Forking regularly by pairs.
*Diclinic.* Of separate sexes; unisexual.
*Dicotyledones.* Having an embryo with two cotyledons.
*Didynamia.* Having two long and two short stamens.
*Diffuse.* Widely and loosely spreading or branched.
*Digitate.* Fingered; spreading like the fingers.
*Digynia.* Two pistils.
*Dimidiate.* Halved, as though one-half were wanting.
*Dimorphous.* Occurring in two forms.
*Diœcia.* Male and female flowers on separate plants.
*Distichous.* Two ranked.
*Divaricate.* Widely diverging.
*Dodecandria.* Having twelve stamens.

*Dorsal.* Relating to the *dorsum* or back.

*Drupe.* A fleshy or pulpy fruit with the seed inclosed in a hard or stony casing.

*Ebracteate.* Without bracts.

*Echinate.* Beset with hooked prickles.

*Elater.* Coiled elastic thread.

*Emarginate.* Notched at the extremity.

*Endosperm.* Layer of substance in the seed usually enveloping the germ.

*Enneandria.* Nine stamens.

*Epicarp.* Outer layer of the pericarp.

*Epigynæ.* Above the ovary.

*Equitant.* Astride, as in leaves of Iris.

*Exocarp.* The outer portion of a pericarp.

*Exserted.* Projecting beyond an envelope.

*Extrorse.* Directed outward.

*Fascicle.* A close bundle or cluster.

*Filament.* The part of the stamen which supports the anther.

*Filiform.* Thread-shaped.

*Fimbriate.* Fringed with narrow cuttings.

*Floret.* A small flower; one of a head.

*Foliate.* Having leaves, leafy.

*Foliolate.* Having leaflets or small leaves.

*Follicle.* A pod formed from a simple pistil dehiscing only along the ventral suture.

*Foveolate.* Marked by minute pits.

*Frond.* The leaf of ferns, liverworts or of Lemna.

*Frustranea.* Florets of the disk hermahprodite, of the ray neutral.

*Frutescent.* Shrnbby.

*Funicle.* The stalk of an ovule or seed.

*Furfuraceous.* Scurfy.

*Galbulus.* An indehiscent cone, becoming berry-like.

*Galeate.* Formed like a helmet.

*Gamopetalæ.* Petals more or less united.

*Gamosepalous.* With more or less united or coalescent sepals.
*Geminate.* Twinned; in pairs.
*Geniculate.* Bent like the knee.
*Gibbous.* Swelling out at one side.
*Glabrous.* Smooth; without hairs or roughness.
*Glaucous.* Bluish-hoary; covered with a fine whitish bloom.
*Glochidiate.* Barbed like a fish-hook.
*Glomerule.* A compact somewhat capitate cyme.
*Glume.* The chaff-like bracts subtending the spikelets.
*-gonal.* Relating to the angles.
*Gymnospermia.* Plants having naked seed.
*Gynandria.* Having the stamens adnate to or borne upon the pistil.
*Gynobase.* A short thick prolongation of the axis upon which the pistil rests.
*Gynœcium.* The pistil or aggregate pistils of a flower.
*Gynophore.* A stem bearing the ovary.

*Hastate.* Triangular or arrow-shaped.
*Heptandria.* Having seven stamens.
*Hermaphrodite.* Both sexes in the same flower.
*Heterosporous.* Bearing more than one kind of spores.
*Hexandria.* Having six stamens.
*Hexagynia.* Having six pistils.
*Hirsute.* Having rather coarse or stiff hairs.
*Hispid.* Beset with rigid or bristly hairs.
*Hyaline.* Transparent or partially so.
*Hypogynous.* At the base of, or below the pistil.

*Icosandria.* Having more than twelve stamens inserted on the calyx.
*Imbricate.* Overlapping
*Imparipinnate.* Pinnate, with an odd terminal leaflet.
*Incised.* Irregularly, sharply and deeply cut.
*Incrassate.* Thickened.
*Incumbent.* Lying upon.

*Induplicate.* With margins folded inward.
*Indusium.* In ferns, the scale-like cover of the fruit spot.
*Inferior.* Lower; below the corolla.
*Inflexed.* Bent or turned abruptly inward.
*Inflorescence.* The flowering portion of a plant.
*Introrse.* Turned inward toward the axis.
*Involucre.* A circle or circles of scales, bracts or leaves, distinct or united, surrounding a flower or flower-cluster.
*Involute.* Rolled inward.

*Labellum.* A lip, as in orchids.
*Labiate.* Lipped; having an irregular calyx or corolla unequally divided into two lips.
*Laciniate.* Cut into narrow slender teeth or lobes.
*Lactescent.* Yielding milky juice.
*Lamina.* The blade or dilated part of a leaf.
*Lanceolate.* Shaped like a lance-head.
*Legume.* A 1-celled capsule opening by two valves, as the pea.
*Lenticular.* Lens, or lentil-shaped.
*Lignescent.* Woody; becoming woody.
*Ligule.* A tongue or strap-shaped body, as the corolla of ray-flowers in Compositæ.
*Lobate.* Divided into lobes.
*Loculicidal.* Dehiscence of the cells of a capsule through the dorsal suture.
*Lomentum. Loment.* A legume jointed, and usually constricted between the seeds.

*Macro.* Large or long.
*Marcescent.* Withering and persistent.
*Marginate.* Furnished with a border.
*Membranaceous.* Thin and translucent.
*Mericarp.* Division of schizocarp.
*–merous.* In parts.
*Micro.* Small.
*Monadelphia.* Stamens united by their filaments into one set.

*Monandria.* One stamen.
*Moniliform.* Like a string of beads.
*Monocotyledones.* Embryo with one cotyledon.
*Monœcia.* Stamens and pistils in separate flowers on the same plant.
*Monogamia.* A syngenetic flower not belonging to a head.
*Monogynia.* Having one pistil.
*Mucro.* A short and small abrupt tip.
*Mucronate.* Having a mucro.
*Mucronulate.* Diminutive of mucronate.
*Multifid.* Cleft into many lobes and segments.
*Muricate.* Rough, with short hard points.

*Navicular.* Boat-shaped.
*Nectaria, Nectary.* Organs which produce a sweet secretion within a flower.
*Necessaria.* Florets of the disk male; of the ray female; or hermaphrodite.
*Nervose.* Having veins or ribs.
*Neutral.* Neither male nor female.

*Obcompressed.* Flattened dorsally.
*Obconic.* Like an inverted cone.
*Obcordate.* Inverted cordate.
*Obovate.* Inverted ovate.
*Obtuse.* Blunt.
*Ochroleucous.* Yellowish white.
*Octandria.* Eight stamens.
*Orthospermæ.* Having the endosperm straight (flat).
*Osseous.* Bony.
*Ovoid.* Egg-shaped.

*Palea, Palet.* Chaff, or a chaffy bract; in grasses the two inner bracts of the flower.
*Palmate.* Compound, with the leaflets radiating.
*Panicle.* A compound raceme.
*Papilionaceous.* Butterfly-like. The peculiar corolla of the Leguminosæ.
*Papillate.* Beset with minute thick projections.

*Pappus.* The hairs, bristles or scales crowning the akenes in Compositæ.

*Parietal.* Relating to or upon the wall of a cavity.

*Paripinnate.* Evenly pinnate; without the odd terminal leaflet.

*Pedicel.* The footstalk of a flower.

*Peduncle.* A general or primary flower-stalk.

*Peltate.* Shield-shaped.

*Pendulous.* Drooping, or hanging nearly inverted.

*Penicillate.* Resembling a brush of fine hairs.

*Pentandria.* Having five stamens.

*Pentagynia.* Having five pistils.

*Perianth.* The floral envelopes—*i. e.*, the calyx and corolla, so far as present.

*Pericarp.* The seed-vessel or ripened ovary.

*Perigonium.* A simple floral envelope.

*Perigynium.* The sac-like envelope, or the bristles or scales which in the Cyperaceæ represent the perianth.

*Personate.* Closed, as in Labiate flowers with prominent palates.

*Petal.* One of the parts of a polypetalous or nearly divided corolla.

*Petiole.* The footstalk of a leaf.

*Petaloid.* Petal-like.

*Phalanges.* Stamens united in sets by their filaments.

*Phanerogamæ.* Plants which fructify by means of stamens and pistils.

*-phyllous.* Leafy, leaved.

*Pilose.* Hairy, with soft distinct hairs.

*Pinnæ.* Primary divisions of a compoundly pinnate leaf.

*Pinnatifid.* Pinnately cleft into opposite nearly equal segments.

*Pinnule.* A secondary pinna—*i. e.*, one of the pinnate divisions of a pinna.

*Pistil.* The female organ of a phanerogam consisting of the ovary with its styles and stigmas.

*Placenta.* That part of the ovary or fruit which bears the ovules and seeds.

*Plicate.* Folded into plaits like a fan.

*Plumose.* Plume-like; having fine hairs on each side, like a feather.

*Polyadelphia.* Having stamens united by the filaments into many sets.

*Polyandria.* Having many stamens.

*Polygamia.* Having both hermaphrodite and unisexual flowers.

*Polygynia.* Having many pistils.

*Pollinia.* Concrete masses of pollen.

*Pomum, Pome.* A fleshy fruit like the apple, inclosing several leathery or bony carpels.

*Procumbent.* Lying upon the ground.

*Pruinose.* Covered with a minute bloom or powder.

*Pseudocarp.* Fruit including other organs in addition to the ovary.

*Puberulent.* Very minutely pubescent.

*Pubescent.* Covered with hairs usually short and soft.

*Pungent.* Terminating in a rigid and stout point or prickle.

*Pyrena.* The stone of a drupe.

*Pyriform.* Pear-shaped.

*-quetrous.* Cornered.

*Quinate.* In fives.

*Raceme.* A form of inflorescence, with pedicellate flowers developing from below upward upon a simple prolonged axis.

*Rachis.* The axis of a spike, etc.

*Radicle.* The part of the embryo below the cotyledon.

*Ray.* The marginal flowers in compositæ.

*Receptacle.* The more or less expanded surface forming a support for the organs of a flower or a head of flowers.

*Reflexed.* Bent abruptly down or backward.

*Reniform* Kidney-shaped.

*Replum.* The frame-like placenta left by the falling away of the valves in Cruciferæ.
*Retrorse.* Turned back or downward.
*Revolute.* With the margin or apex rolled backward.
*Rhizome.* An underground, somewhat horizontal, rooting stem.
*Ringent.* Gaping; applied to labiate flowers with open throats.
*Rostrate.* Beaked.
*Rotate.* Wheel-shaped.
*Rugose.* Wrinkled; ridged.
*Runcinate.* Deeply-toothed or lobed, with the segments directed backwards.

*Saccate.* Sac-shaped.
*Samara.* An indehiscent, membraneously-winged fruit, as in the ash and maple.
*Sagittate.* Shaped like an arrow-head
*Scabrous.* Rough to the touch, with minute rigid points.
*Scape.* A naked peduncle rising from the ground.
*Scapigerous.* Producing scapes.
*Scarious.* Thin, dry and membranaceous.
*Schizocarp.* An ovary which divides into its component parts, each part usually remaining closed over its seed.
*Secund.* One-sided.
*Septicidal* Opening through the lines of junction of the carpels.
*Septifragal.* Breaking away from the partitions in dehiscence.
*Serrate.* Having teeth directed forward, like those of a saw.
*Serrulate.* Finely serrate.
*Sessile.* Without footstalk.
*Setaceous.* Bristle-like.
*Siliculosa.* Short cruciferous pod, as Capsella.
*Siliquosa.* Long cruciferous pod, as in mustard.
*Sinuate.* With a strongly wavy margin.

*Sinus.* An indentation, either angular or rounded, separating parts.

*Sordid.* Rusty or dirty-colored.

*Sorus, Sori.* A cluster of sporangia.

*Spadix.* A spike with a thickened, fleshy rachis, usually subtended by a spathe.

*Spathe.* A sheathing bract.

*Spathulate.* Narrowly attenuate downward from an abruptly rounded summit.

*Spicate.* Like a spike.

*Spike.* Resembling a raceme, but the flowers sessile, or nearly so.

*Spinescent.* Ending in a spine or spiny point.

*Spinulose.* Having diminutive spines.

*Sporangia.* Cases which contain spores.

*Spores.* The minute bodies in cryptogams, which answer somewhat to the seeds of other plants.

*Squamose.* Furnished with scales.

*Squarrose.* Irregularly spreading.

*Stipitate.* Borne upon a stipe.

*Stipule.* An appendage to the base of a petiole.

*Stamen.* The male organ of flowering plants, consisting of an anther borne usually upon a filament and containing the pollen.

*Staminodia.* A sterile or undeveloped stamen.

*Strobilaceous.* Inflorescence formed of imbricated scales, as in the Coniferæ.

*Subulate.* Awl-shaped.

*Suffrutescent.* Slightly shrubby; woody at base.

*Sulcate.* Grooved or furrowed.

*Suture.* A line of union.

*Synandræ.* Having united anthers.

*Syncarpous.* Having united carpidia.

*Syngenesia.* United anthers.

*Tendril.* A thread-like production in climbers.

*Terete.* Cylindrical, or nearly so.

*Ternate.* In threes.

*Testa.* The outer seed-coat.
*Tetradynamia.* Having four long and two short stamens; Cruciferæ.
*Tetragynia.* Having four pistils.
*Tetrandria.* Having four anthers.
*Thalamus.* End of peduncle.
*Thallus.* A cellular expansion taking the place of stem and foliage.
*Thyrse.* A contracted or close ovate cyme.
*Tomentose.* Pubescent, with matted wool.
*Triandria.* Having three stamens.
*Trigynia.* Having three pistils.
*Triquetrous.* Triangular, with the sides concave or channeled.
*Truncate.* Ending abruptly as if cut off transversely.
*Tumid.* Swelled.
*Turbinate.* Top-shaped.
*Turgid.* Inflated.

*Umbel.* An umbrella-shaped inflorescence, the pedicels radiating from the summit of the common peduncle.
*Umbilicate.* Pitted in the centre.
*Uncinate.* Hooked at the extremity.
*Undulate.* Wavy.
*Unguiculate.* Narrowed below into a claw.
*Urceolate.* Cylindrical or ovoid, but contracted at or below the open orifice.
*Utricle.* A small, bladdery, usually 1-seeded, pericarp.

*Vallecula.* Groove between the ribs of umbelliferous fruits.
*Valvate.* Opening by valves.
*Ventral.* Belonging to the anterior or inner face of a carpel; the opposite of dorsal.
*Ventricose.* Swelling unequally or on one side.
*Versatile.* Swinging; turning freely on its support.
*Verticillate.* Arranged in whorls.

*Vexillum.* The large upper petal of a papilionaceous flower.

*Villous.* Bearing long and soft straight, or straightish, hairs.

*Virgate.* Slender, straight and erect.

*Vittæ.* The longitudinal oil-tubes in the pericarp of most umbelliferæ.

*Whorl.* An arrangement of leaves, flowers, etc., around the stem.

# ERRATA.

Page 106. After generic description of Rigiopappus read "R. leptocladus.—Niles. Spring."

Page 163. For "Gillia" read "Gilia."

Page 163. For "androscaea" read "androsacea."

Page 171. 6th line from bottom, for "decuscate" read "decussate."

Page 172. 9th line from bottom, for "L" read "S."

Page 215. 7th line from top, after "♄." insert "—Niles. Summer."

Page 262. 10th line from bottom, omit comma between "late" and "dehiscent."

Page 275. 12th line from top, for "7" read "5."

Page 283. 4th line from top, for "Californicus" read "Californicum."

Page 299. 11th line from top, after "not" insert "at."

Page 311. 1st line, for "tenela" read "tenella."

Page 330. For "Pillularia" read "Pilularia."

www.ingramcontent.com/pod-product-compliance
Lightning Source LLC
Chambersburg PA
CBHW030404230426
43664CB00007BB/745